being of the sun

by ramón sender barayón & alicia bay laurel

drawn & lettered by alicia, with musical notation by ramón. songs: music by ramón & lyrics by alicia, except where otherwise noted

EPBM echo point books & media, llc
brattleboro, vermont

☆ published by Echo Point Books & Media ☆

Brattleboro, Vermont

www.EchoPointBooks.com

being of the Sun ISBN 978-1-63561-985-0

https://raysender.com

https://aliciabaylaurel.com

This is the 2nd edition. The first edition was published in 1973 by Harper & Row, New York.

interior design, lettering, illustrations, cover design and cover art by Alicia Bay Laurel

acknowledgments

Huge thanks to Marshall Glickman, owner/publisher at EchoPoint Books & Media, for making the decision to publish this book—the first new English edition since 1973!

Great gratitude to Fred Lee, production manager at Echo Point, for guiding the process from layouts to printed copies.

Thank you so much, Anneka Kindler, Contracts & Royalties real human being at Echo Point, for working on our contract with Alicia.

Enormous thanks to fine artist and digital designer Karen Tsugawa, who coached Alicia over Zoom in assembling (via PhotoShop 2020) the layouts requiring revision.

Deep bow to author Gregory Sams for the elegant foreword.

And sincere thanks to all of the readers over the past 45 years that encouraged us to get our book back in print again!

please feel free to color in these pages. no one has to color inside the lines. that way each book is unique.

contents

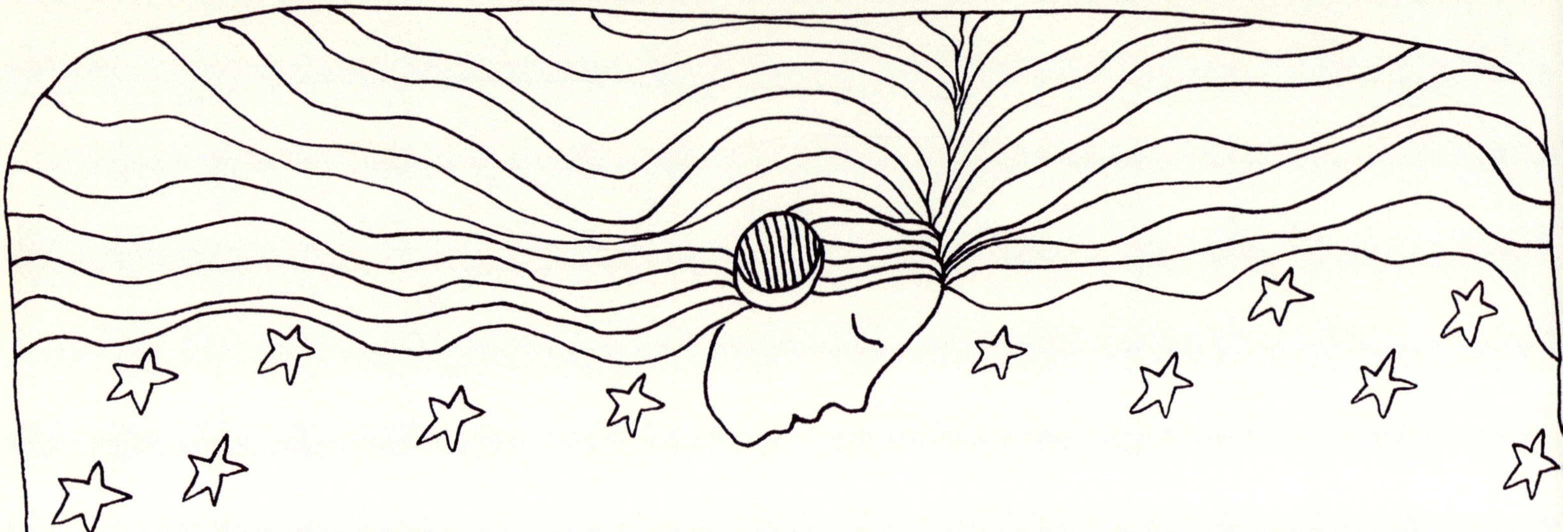

For me, creating this book in collaboration with my dear brother & mentor, Ramón Sender Barayón, was an expression of soul freedom. I asserted my veneration for the holiness of nature, on earth, in space, in the entity through which I manifest, in the intersections of time & life cycles, as the sources of light & shadow, the wheels of the seasons, the hours, & the phases of the moon. I declared the freedom of every being to find a unique path to experiencing oneness with all that is, with or without reference to the thoughts of previous seekers.

I have returned to remember this freedom, and to treasure it.

Alicia bay laurel ~ Summer 2020

In writing this book, we came to listen more deeply to Aditi's child, who is the source of all life on our planet, and one of her uncountable children, the Adityas who brighten our galaxy. Meditating on our creator star during the 1960s, I realized that our dualistic view of the universe was a misconception. Matter, mind & spirit are just different vibrations- different notes on the same scale. And, if the Sun is a conscious being, then everything else must be conscious as well.

All is one, and that unity does not mean that only a material plane exists. Spirit & matter intermingle in ascending & descending cosmic melodies.

Ramón Sender Barayón ~ Summer 2020

the glorious Sun–
it powers our being and
brings life to every plant &
creature on this beautiful planet
Without it, we would not be.
What could be more deserving of
gratitude than our local star,
acting as a conduit for the divine
light of creation itself?

Our spirits are lifted when we bask in Sun's light, and plants grow from tiny seeds to beautiful flowers & towering trees. We recycle the vibrations of our local star into the energy of life.

For millennia, it was our default mode to regard Sun as a divine being. Though it stars in the movie of life, many today give more thought to sunglasses than Sun itself, seeing it as no more than an accidental light bulb in the sky.

Humanity
can benefit
beyond measure
by bringing Sun back
into our lives, enriching
our hearts and spirits
with its infinite love.

Having written a book on
the subject myself, I was
delighted to discover this
beautiful love story, written by
Alicia and Ramón over thirty
years earlier.

Now resurrected, their sumptuous
pages connect us to our local
star in words and pictures
infused with a love born of
genuine communion. The
authors have been there, and
share the bliss.

gregory Sams, author, Sun of gOd
www.gregorysams.com

we dedicate this book
to our dear father Sun
source of the earth's children
may your voice sing in every
cell of our bodies and shine
ever more from our hearts.
may we be mirrors of your light.
we love you
with your own love
O our god star

• the best way to get high is the one you discover yourself.
The way you discover yourself is to find a way to get high

one cosmic experience of your own is worth 10,000 second-hand ones. In american indian cultures people used to have visions which became their personal religions or ways of getting high. a vision expands one's awareness and makes living easier and more enjoyable. Many beings have expanded their awareness to blissful states of at-One-ment with the universe. Some have taught other people their methods & beliefs. many have followed these divine inspirations, but one may discover one's own way, as did these teachers.

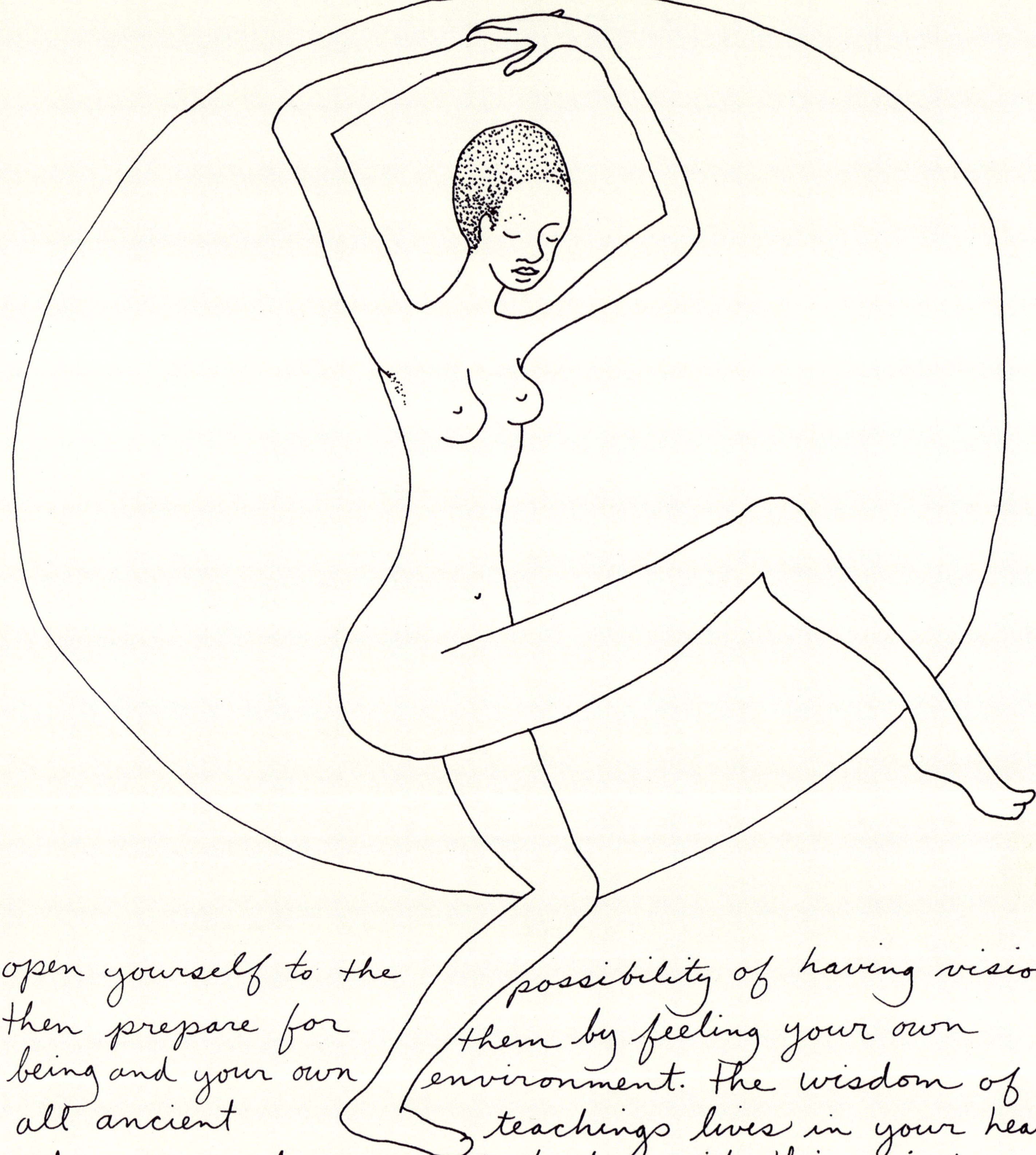

open yourself to the possibility of having visions. then prepare for them by feeling your own being and your own environment. the wisdom of all ancient teachings lives in your heart. when you relax enough to hear it, this wisdom can rename you, reclothe you, give you dances, exercises & meditations, ceremonies & recognitions of divinity in everyday life that make your whole day an act of being radiantly blissful.

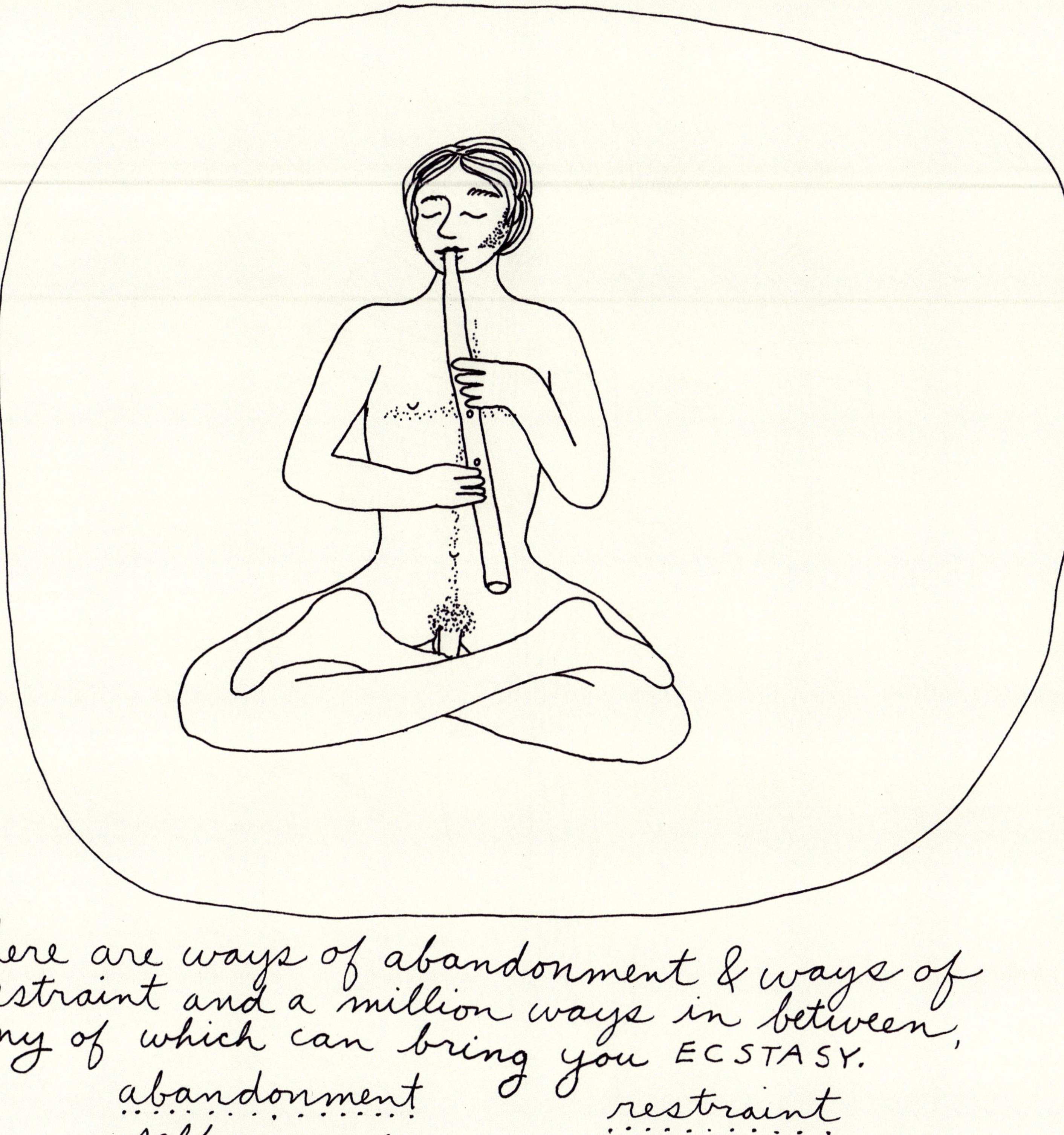

there are ways of abandonment & ways of restraint and a million ways in between, any of which can bring you ECSTASY.

abandonment	restraint
self-expression	rules of behavior
ingestion of a sacrament	study & meditation
improvisational dancing	exercises & rituals
feasting	dietary laws
sense awareness	tuning out of senses
wilderness	temple or monastery
nudity & costumes	prescribed dress
love-making	chastity

your friends can teach you ways to make your life happier and induce bliss states. they can teach you by example; you need not express to the person that they are your teacher, nor need you limit yourself to one. If you spend a night in the same house your souls will merge as you sleep. If you meet someone who is admirable to you in every way, including life style, you might surrender to him or her that they may teach you their way. the qualities you admire in them cannot be recognized unless they are resonating in you at the same time. the teacher is a reminder of your highest self, and in that sense, your guru is you. animals & trees are also great teachers. mountains, wind, ocean and sun are teachers. lessons abound for the observant.

every religion was born in one time & place. they do not travel well through time & space. at their best, each religion applies to one person, to a family or to a tribe.

to cling to the past is to cherish abstractions. drawing from experience of the here & now is the way to find your own religion.

the rest of this book is an example of a homegrown personal religion, which we have been discovering in the here & now of our lives and from teachings which have inspired us. these beliefs and methods of getting high have been discovered in the here & now of the lives of people we have met, so we do not claim them as ours alone. a consciousness is manifesting on the planet in many people at the same time. sun yoga is as old as the sun.

any religion is a way
any way to get high is a religion.
any religion is a way to get high.
to get high. grow your own..

what we believe

the earth was made from
one of the Sun's ribs.
the earth is all of our mother and
the sun is all of our father
all life comes from the divine
interaction of sun & earth.
spirit & matter
can never be separated.
we are the children
of their eternal passion.

as sun-spirits we come into human form until we prove harmless to the planet. then we can graduate to being something closer to the earth, such as an animal or a tree.

the sun is our model of a loving consciousness. enlightenment is a sunny disposition. the more you relax, the more loving your consciousness becomes, until it shines through all acquired conditioning and you are as radiant as the sun.

gazing at sunlight is relaxing & quieting to the mind. living in nature, away from civilization, in a sunny place is the simple road to bliss. (see warning on page 200) life is lived with ease & grace when the self rises above the mental chatter to its true abode. a great rush of sensory awareness occurs as you enter the here & now and you discover its many levels: the here & now of the cells of your body, of everything within your

view & hearing, of your continent, hemisphere, the whole earth (& inside her), the whole solar system. you become telepathic with these beings. the here & now of the sun is $7\frac{1}{2}$ minutes in our future (the time it takes sunlight to reach the earth). living thus in the future, your movements might synchronize with the wind or the flight of birds. cause & effect merge. All is energy.

the union with solar consciousness allows you to live in the community of the stars, who are, after all, just you in your infinite colors, forms, disguises, nebulae: the endless playground of the nameless One. your body, transformed into light-energy and needing neither air nor food, can travel anywhere in the solar system.

a light body has been described in many religious writings. we like to call it a rainbow body because its aura contains all the colors of the spectrum.

the earth has a rainbow body as her atmosphere.

O SUN

creator & source of all
terrestrial life ... we developed eyes
to witness your bounteous light.
fish leap out of water into you
and crawled out onto the land
to follow your gaze. So does
the lizard love to bathe in your
light, you are his mobility.
earth life greets you rising in the east
and calls after you as you disappear;
roosters crow & cows moo with
the coming & going of your great light.
birds fly to the tree tops to sing
their evening songs in your last rays.
they follow you as our mother tilts north & south.
bees always look at you & fly toward you.
sunflowers, morning glories, all green leaves
turn to you as the Source.
the whole planet yearns to be One
with you through her creatures.
let a new golden age begin for all
of your children and
reign forever!

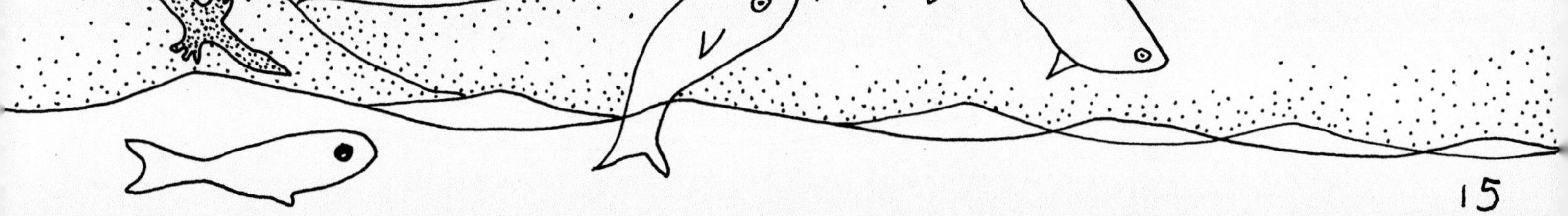

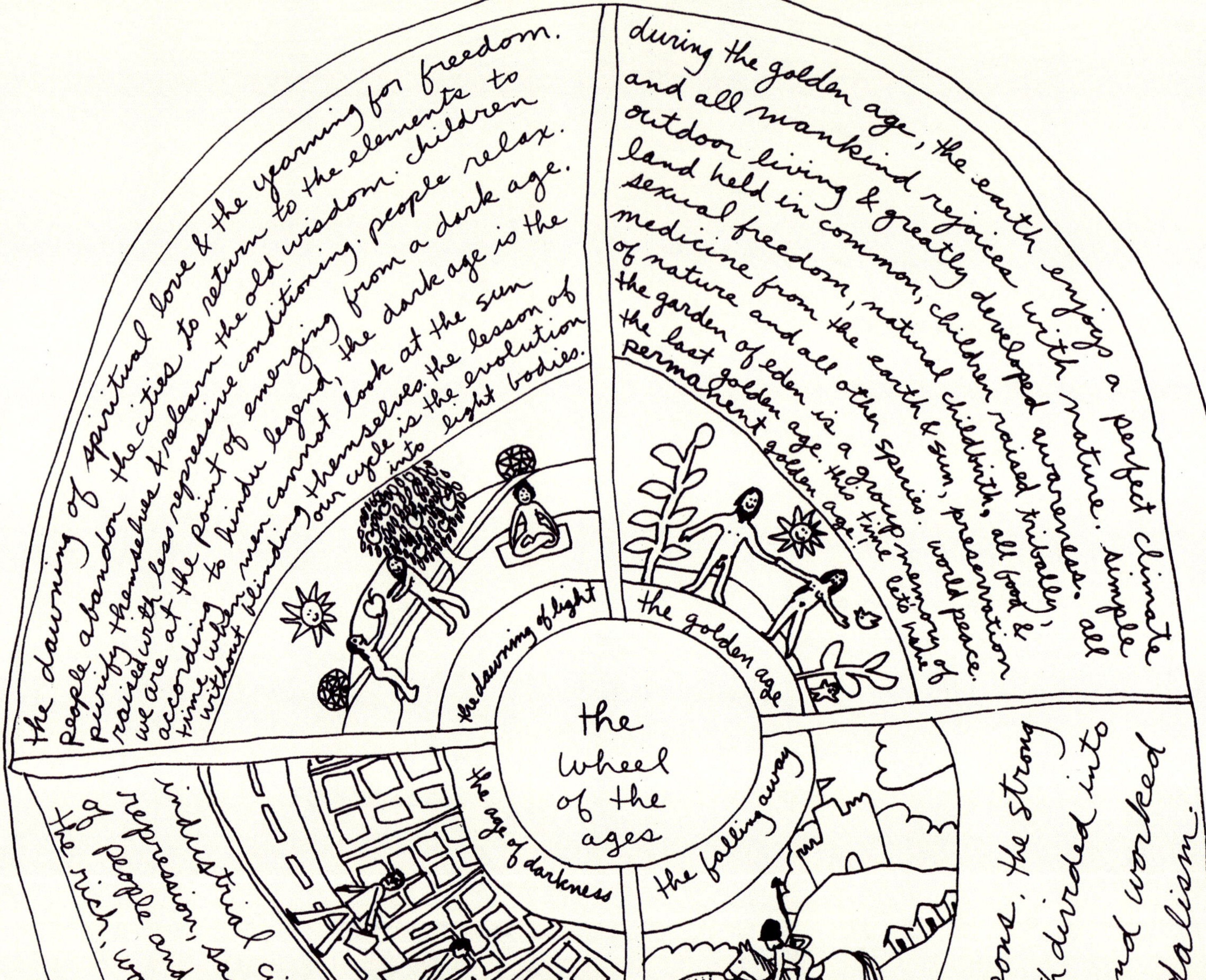
the dawning of spiritual love & the yearning for freedom.
people abandon the cities to return to the elements to
purify themselves & relearn the old wisdom. children
raised with less repressive conditioning. people relax
we are at the point of emerging from a dark age.
according to hindu legend, the dark age is the
time when men cannot look at the sun
without finding themselves. the lesson of
our cycle is the evolution
into light bodies.
during the golden age, the earth enjoys a perfect climate
and all mankind rejoices with nature. simple
outdoor living & greatly developed awareness. all
land held in common; children raised tribally;
sexual freedom; natural childbirth; all food &
medicine from the earth & sun; preservation
of nature and all other species. world peace.
the garden of eden is a group memory of
the last golden age. this time into the
permanent golden age!
the development of weapons. the strong
enslave the weak. the earth divided into
properties. large holdings of land worked
by the poor to benefit the rich. feudalism.
disease, lack of basic freedoms.
development of monetary
systems: civilization
industrial cities, armies, wars, sexual
repression, sadism, pollution, wanton killing
of people and other animals, oppression of poor by
the rich, women by men, nature by technologists,
children by adults, and all
by rampant loneliness and
ill health.
the dawning of light
the golden age
the falling away
the age of darkness
the wheel of the ages

soul mates

the sun is a great arranger of marriages because all love emanates from its light. when you feel the sun as your lover, the sun sends you its love manifested in the person of your true love, in whose being the images of all beings you have ever loved exist harmoniously. as you share more & more deeply your here & now, you become telepathic and closer to solar consciousness. you become the divine interplay of the sun and the earth.

The Four Liberating

(a yoga is a path to union with the universe)

earth yoga: generosity. making free land, food & health services available to all. ecological land use. raising all your own food & medicine. planting trees. raising earthworms. returning organic garbage & shit to the soil instead of polluting rivers and oceans. reverent use (not littering) and opening to all—the holy places of the earth (beaches, lakes, mountain tops, hot springs, volcanic craters, etc). picking up litter. building your home & living to blend with nature & not impose upon her. protecting wild life. protecting forests. perfecting your health.

examples of earth yogis: the diggers, open land communes, ecology action groups, the hog farm, sierra club.

water yoga: love, flowing with life, speaking & acting honestly, making love, crying, laughing, hugging, kissing, touching, feeling, sensitivity to the needs of others, kindness, gentleness, sensuality, sense awareness, extended families & other new relationships, caring for the sick & unhappy. water yogis: esalen groups, communal families.

Elemental yogas

air yoga: truth.

telepathic communication. psychic readings. astral travel. non-repressive schools for children. student-run universities. freedom of speech in all media: truthful news reporting, underground publishers, open radio, TV, theatre, galleries. freedom of expression instead of commercial ventures. information of all government activity open to all. information of earth resources open to all. promoting world peace. tuition-free trade schools. unifying the whole earth as one people with equal freedoms & no leaders. air yogis: the world game, free universities, free schools, whole earth catalogue, non-commercial media

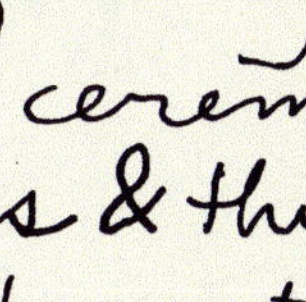

fire yoga: consciousness.

expanding awareness. relaxing. rising above the thought process. meditation. mantra. tuning in to the sun. finding the divine within. prayer. sacred ceremonies. making of holy times & places & things. all who are dedicated to getting high are fire yogis.

you don't have to be a bird
to live in a tree.

sunlight
meditations

"if I am asked whether it is my nature to revere the Sun, I say again: it is! for the Sun, too, is a manifestation of the highest, and is indeed the mightiest that we children of earth are privileged to look upon.

I worship in it the light & the creative power of God, through which alone we move and have our being, and all plants & animals with us."

goethe's conversations with eckermann

everything is better out in the sunshine. meditating upon sunlight is the simplest way to get high; very pleasant sensations fill your solar plexus. The mind is cured of thought and the body of sickness. lucidity & awareness & comprehension are increased by sunlight, so learning is easier outdoors. making love in the sunshine is the greatest freedom. balance exposure to direct sunlight with variations of shade; overdosage of certain rays can damage your skin, your eyes or your hair. we won't take responsibility for anyone neglecting their bodies & blinding themselves. go slowly, gently & gradually change.

(see warning on page 200)

sunrise & sunset gazing

the sun's rays coming through the earth's rainbow aura make the colored light that plays upon the clouds, haze & even the earth herself. the band of color that the light is shining through determines what color they reflect. the pink tones are easiest to see in the blue, white & grey of the sky, but with relaxed observation, all colors can be seen.

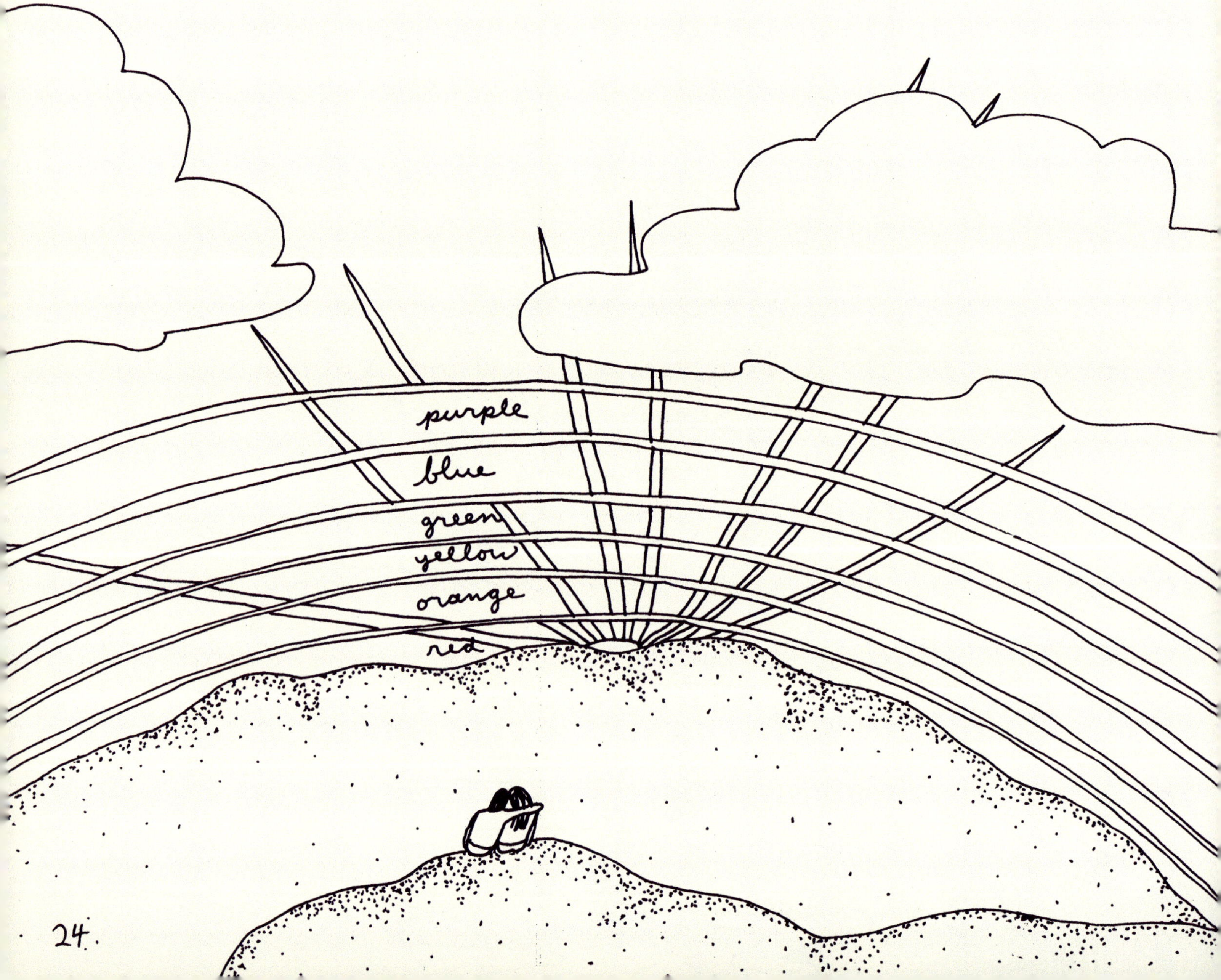

rainbows & other wonders

when the sun is near the horizon, shining through falling rain, on the opposite horizon appears a refraction, a rainbow. sometimes when you are flying in a clear space between clouds, the upper cloud will refract the sunlight on the lower cloud as a circular rainbow with your shadow in the center. when the sun is just behind a thunderhead, sometimes it colors the surrounding vapors into auras iridescent as oil floating on water. Patient observation is rewarded with lightning, shooting stars, and all the celebrations of nature.

earth meditation a mountain valley is a temple for contemplating the earth. a greenhouse makes a wonderful rainy-day meditation room. the plants can be part of your ceremony and will flourish from your attention & good vibrations.

sky meditation a hilltop is a sky temple. tranquil contemplation can be found in the colors of the sky & the movements of clouds, as you lie on your back or do hatha yoga.

water meditations

the motion of reflected images on water remind us of the impermanence of things. to gaze upon them is soothing; to float on water is to be rocked by the great mother. bathing & swimming cleanse the mind as well as the skin, weightlessness affects the whole being. the reflections of sunlight on water make a shimmering light beautiful to look upon. the water casts lines of moving light upon cliffs, trees & walls around it. dappled shadows of overhanging leaves make dancing shapes with the movement of water. waterfalls make rainbows in sunlight. moving water produces an endless variety of drone sounds. ocean surf makes a fantastic light show, especially when viewed from above, from a cliff or a boat.

trees as sun temples

redwood, coconut palm, quaking aspen, silver birch, each tree has its own way of filtering sunlight. each leaf & branch performs a different dance with the wind; each tree has a unique voice that sings with the mother's breath (as do we). each grove is a choir that raises its voice in harmony when moved by God. temples must have evolved from sacred groves. when a tree died, it was replaced by a pillar. temples are groves in countries losing their topsoil, like greece and egypt. (we like to feed our topsoil and groove in groves.)

if too many people visit a sacred grove, it soon becomes a temple because the trees get cut down for souvenirs and directional signs.

orchards are temples to the earth's bounty and the sun's light. by eating the fruit and burying your shit deep in the ground (leave no tracks), you plant the seeds of groves to come. even a single tree can be a sacred sun temple. lying under trees & looking up at the sunlight coming through

the leaves is one of our favorite sun meditations. a book of blank pages is good to take along for an afternoon under a tree. the sunlight makes patterns you can trace. a hammock makes a comfortable place to sit; a tire hung by a rope makes a sunstrobing seat (see sunstrobes). just be sure your tree is strong enough not to be bent or broken by your weight.

our wedding ceremony took place in a temple made of a tall pine tree and a little oak that grew in a passionate embrace.

plant a sacred grove

choose trees that are your favorites (ours are redwood & bay laurel). plant trees whose shade you like or whose fruit you like or whose smell you like. find out what kind of soil, compost, drainage, amount of water, and climate are most enjoyed by the species of tree and arrange your chosen site in accordance. Study how the sun moves across the land. plant the trees in a circle or some other holy shape. some early sun symbols:

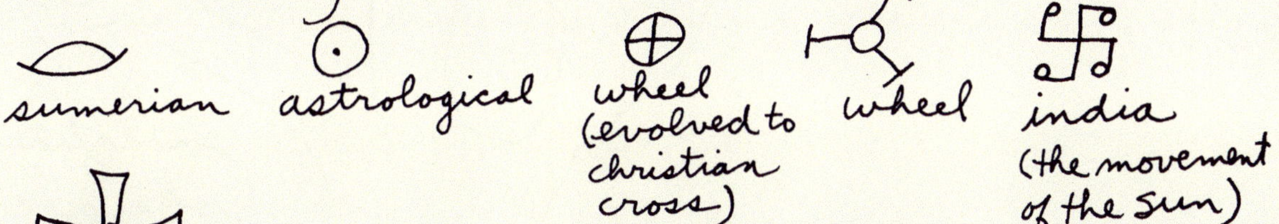

leave room between trees to allow for maturity. dig a hole deep enough for the ball of roots & the compost below it. (you may take a sacramental shit under each tree). give the young tree plenty of water. you may have to fence them off until they are large enough to withstand the company of grazing animals. plant flowers around the grove.

sing to your trees at least once a week. encourage birds to live in your grove. strive to be like the trees. play music near them. they're nice to have as friends.

to be perfectly enlightened is to be like a tree: firmly rooted in the earth, constantly adoring the sun, drawing energy from sunlight, improving the air (consciousness) around it, improving the soil both before & after death, passive to death, singing when moved to do so by the wind passing through its body.

how to join a grove of trees: hang out with them a lot. learn their language. hug them, lie next to one with the soles of your feet on the trunk. You will feel

the wind in the branches. When you are living in harmony with their ways, they will accept you as a junior member of the grove & initiate you as a novice tree.

laurel

there was once a nymph named daphne who loved the sun so much that she became totally beautiful to the sun. but in a human body she could not give herself totally, so she changed into a being that lives on sunlight, a laurel tree. The leaves of the laurel were burned on braziers at delphi so that the sibyls could inhale the smoke and enter the deep trance in which they foretold the future. The laurel is medicine, like sunlight; as a tea against colds or placed over the steaming rocks in a sauna bath. a crown of the leaves keeps head lice away, rolling in a forest floor of bay will rid a dog of fleas.

redwood.
great dark quiet red-
wood trees grow in circles.
the parent tree starts young
trees around it; by the time it
dies there is a circle of tall
trees. such was the sun temple
where I first began meditating
on sunlight coming through
the branches, while lying on the
fragrant needle floor, breathing
the abundant oxygen. I talked
with them. they are among the
largest & long-lived trees; they
are patient & talk very
slowly. It's good to live
around long-lived
creatures when
learning to slow
down

warning: strobed light can induce seizures in people prone to epilepsy. if your muscles contract while using a strobe, stop at once.

sun strobes are devices for alternating direct sunlight with darkness, a light & shade pulse to be viewed through closed eyelids. the strobing produces colored mandalas playing on the inside of the eyelids which sometimes flash messages to you &, when very intense, produce a blissful white light state.

the simplest sunstrobe is your hands. spread the fingers of one hand and wave between your closed eyes & the sun. an opaque pin wheel works if the blades are long enough to shadow both eyes. an old black umbrella with every other panel cut out, twirled between you & the sun, works well. the world is full of strobes: windmills, electric fans, roof ventilators, running by a slat fence with the sun behind it, or riding down a tree-lined street.

sunstrobes you can build

this is a 2-person model; one strober, one viewer. cut out a circular cardboard disc about 12 inches in diameter. Cut out two slots, each 1/8 of the circumference or the distance between the outer corners of your eyes.

mount it on the tip of a wooden dowel or a straight branch several inches long with a pushpin or a nail so that it turns on the shaft of the nail. Glue buttons near center to give a place to push with your fingers as you twirl it over the viewer's closed eyes as they face up toward the full sunlight. ....

battery-operated sunstrobe

made of a small battery-operable motor (with shaft plus mount):

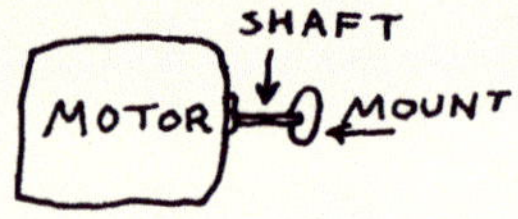

one low-resistance potentiometer
one ½" wooden dowel
one C (flashlight) battery
a few feet of coated hook-up wire
a foot of stiff baling wire
liquid or hot solder
a copper band
mounting cement.

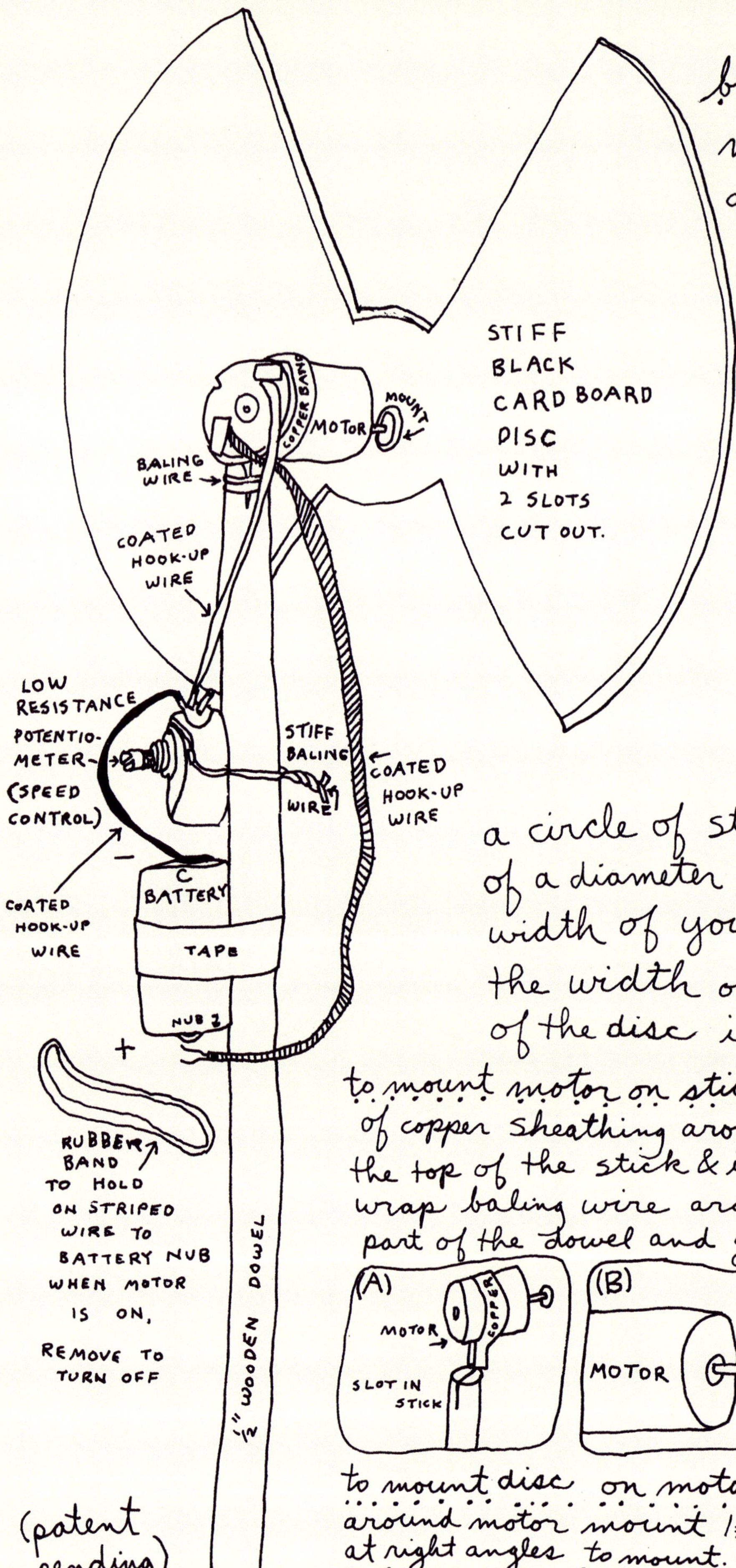

a circle of stiff black cardboard of a diameter larger than twice the width of your face. (draw with a compass)

the width of the slot at the outer edge of the disc is the width of your face.

to mount motor on stick (A) wrap a one-inch-wide strip of copper sheathing around the motor. cut a slot in the top of the stick & insert ends of copper strip, then wrap baling wire around the outside of the slotted part of the dowel and glue with mounting cement.

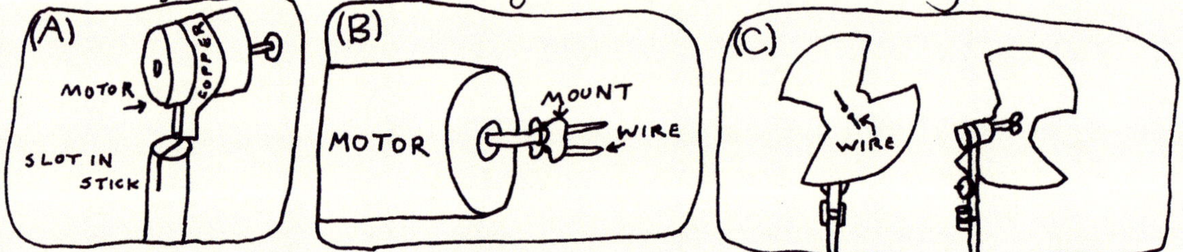

to mount disc on motor mount (B) wrap stiff wire around motor mount 1½ times. cut off 1½" long and bend at right angles to mount. center on disc, mark & puncture holes in disc & insert wires. push disc tightly against mount and bend wires against disc pointing towards outer edge of disc. glue over the wire on mount and on disc (C).

(patent pending)

wire the potentiometer to the dowel with baling wire and mount the battery to the dowel with heavy tape or light wire.

wire with hook-up wire one motor terminal to the center terminal of the potentiometer. attach another wire to the other motor terminal long enough to reach the positive + nub on the battery. wire either of the outside potentiometer terminals to the negative − end of the battery using tape to attach wire to battery. use hot or cold solder to attach wires to motor terminals & potentiometer. To turn it on, attach motor wire to positive + nub of battery with a rubber band.

for indoor use, glue a centering spiral to the center of the disc so that when the motor turns, the spiral appears to disappear inward. this makes an effective mandala for meditation and is sometimes used for inducing hypnotic states.

solar-powered sunstrobe

instead of a battery, mount with glue on the end of the motor a photo-sensitive cell. (there are special small motors & super low resistance potentiometers made to run off of them.) the difference in hook-up wiring: wire one of the wires of the photo-sensitive cell to the motor terminal and the other wire to one of the potentiometer terminals. wire the other motor terminal to the potentiometer and solder ends in place. sun turns it on. (D)

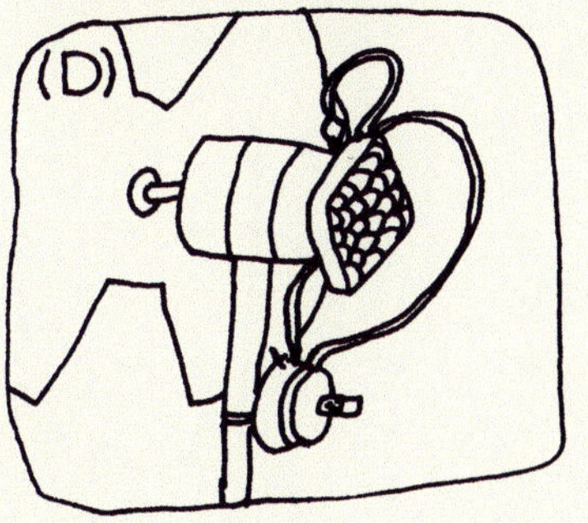

other motors

in secondhand stores you can sometimes find battery-operated appliances that will turn the black slotted disc; shoe buffers, car vacuum cleaners, desk-top fans could be wired to be sunstrobes.

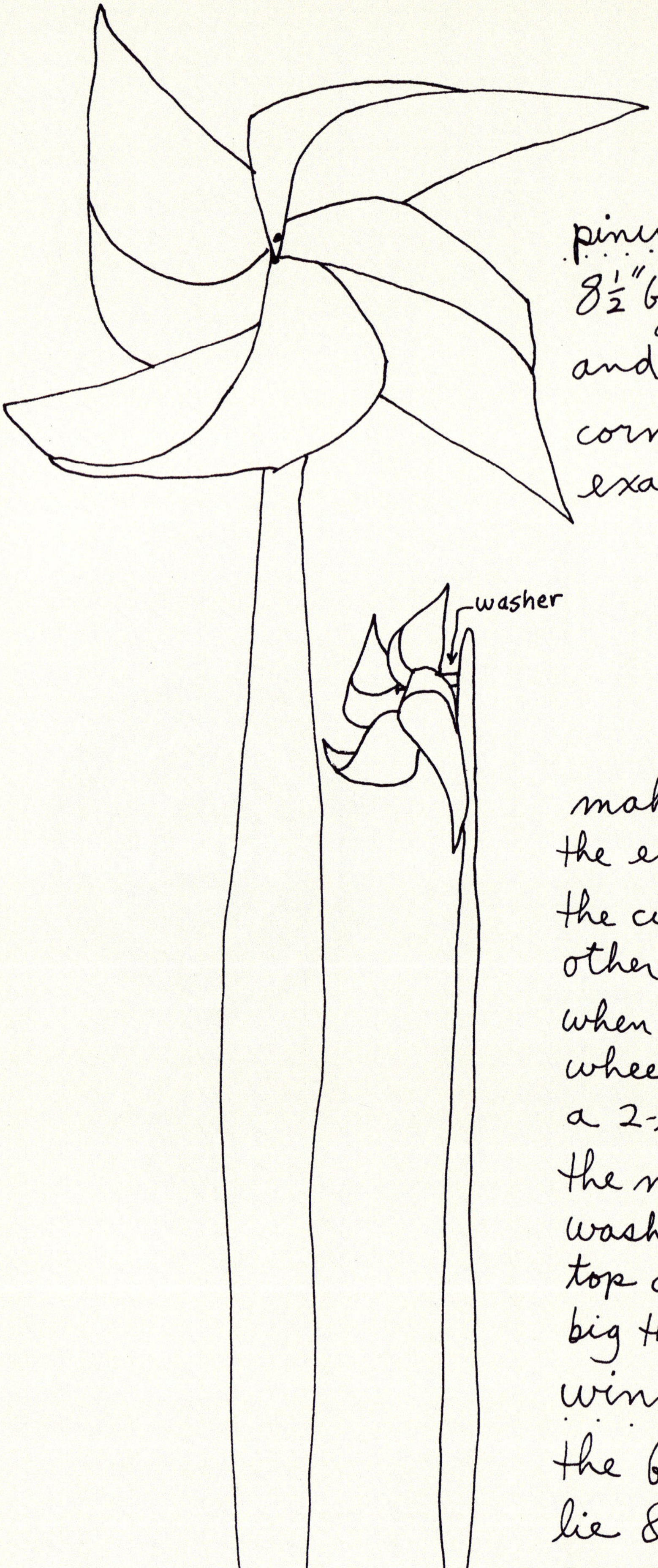

wind-powered sunstrobes

pinwheel: cut out a square $8\frac{1}{2}$" by $8\frac{1}{2}$" of heavy black paper and cut from each of the corners to $\frac{3}{4}$" away from the exact center of the square.

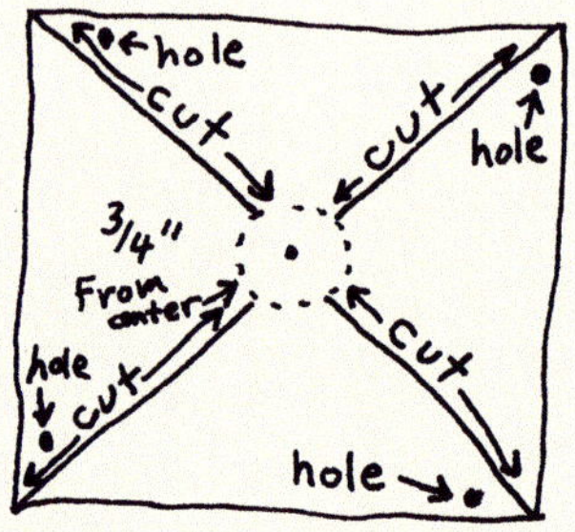

glue a cardboard circle $1\frac{1}{2}$" in diameter over the center of the square. make a hole with a nail through the exact center of the square & the circle. make holes in every other corner of the sails (above). when the glue is dry on the center wheel, thread all the holes onto a 2-inch flathead nail and pass the nail through the center. make a washer of a $\frac{1}{2}$" of thin plastic tube or a top of a ballpoint pen. nail to a big thick wooden stick or a broomstick.

windmills: in the shadow of the blades of windmills you can lie & observe the strobing, but sometimes it's hard to get together the directions of sun & wind.

our dream suntemple sunstrobe a circular structure with the ability to strobe from the center out any amount of the roof, which can tilt towards the ecliptic (the sun's path). the strobe would have speed control and the roof would be movable to control the amount of sunlight that could enter the room. The floor would be comfortably padded so people could lie on their backs with their eyes closed and meditate on the brightly colored mandalas inside their eyelids. Any inventor-architect with a feasible design, please let us know!

how to use a sunstrobe (see warning on page 200)

lie on your back, or sit with your head back, eyes closed and centered up in the forehead. keep the strobing on both eyelids. keep the strobe close enough to eyelids to block peripheral light but don't bump your face with the disc! (for children's use the vanes could be made of cloth so that they flop over when the motor isn't turning.) try different tensions of eye muscles (squeezing & relaxing.)

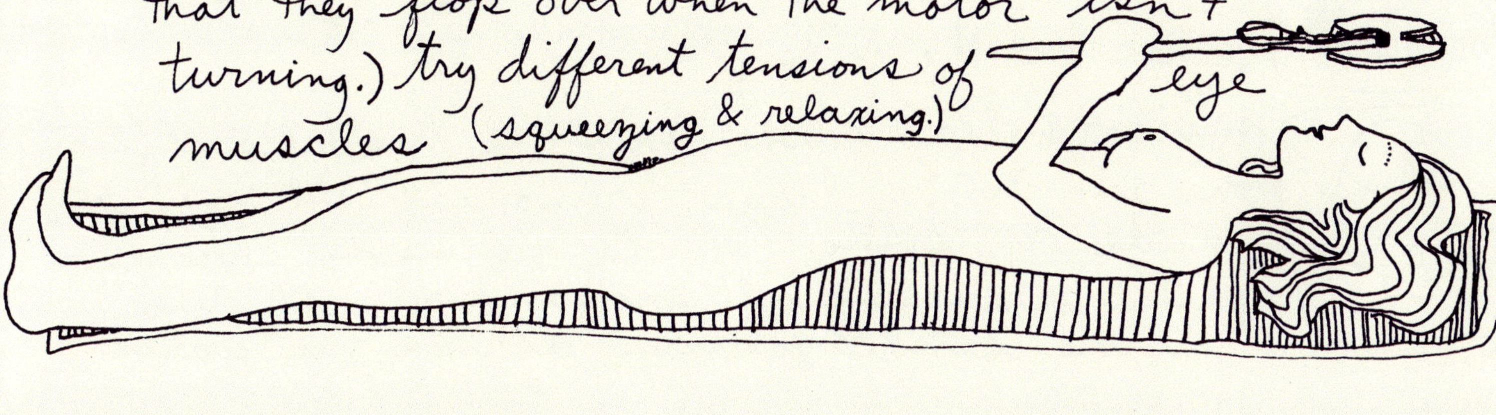

solar light shows

rainbow makers (refractors)

hang in a sunny place some big crystals from a chandelier. or place a fish bowl or aquarium full of water in a sunny window. wet eyelashes make rainbows. four sheets of clear plexiglas assembled into a standing triangular prism filled with water (glue with epoxy):

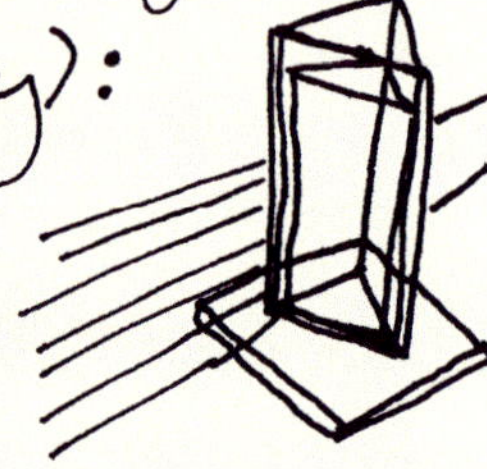

reflectors

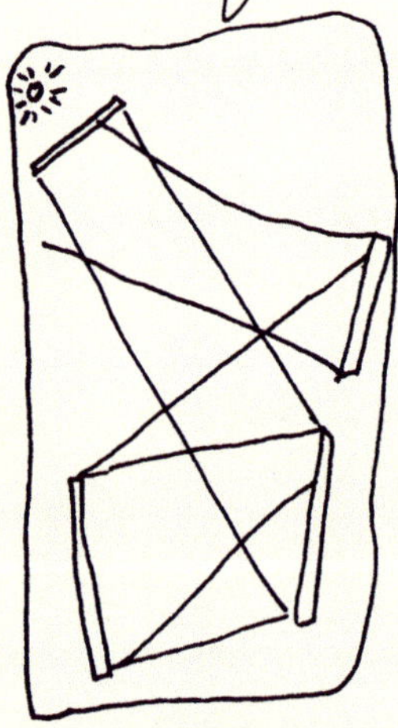

mirror light sculpture: place a mirror where it reflects the sun onto an opposing wall. place another mirror where the sun hits the opposing wall, bouncing the light onto still another wall. place another mirror on this reflection and keep adding mirrors onto reflections. then light a stick of incense and dance in the middle of all the bouncing light.

mobile: make a mobile with pieces of mylar (mirror plastic) and pieces of colored plastic gels (and maybe a prism or two).

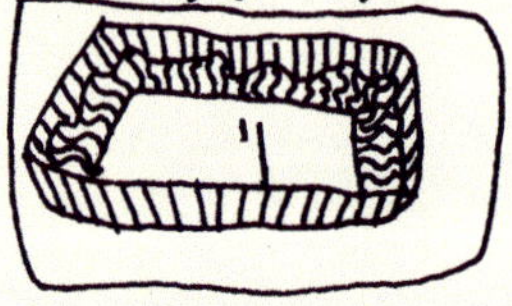

water: place a mirror at the bottom of a pan of water on a window ledge to reflect moving lines on the ceiling.

mirror balls: silver christmas tree balls reflect the sun 360°. a large ceremonial ball could be made by having an old bowling ball chrome-plated. a turning ball mosaicked with tiny mirrors sends galaxies across the walls.

filters: stained glass windows, bottles filled with water (each with a different color of food coloring in it), tent of translucent patchwork (parachute rip-stop nylon scraps: first glue together, when dry, sew together). ping pong ball, cut into hemispheres - place one over each open eye (for meditating on sunlight).

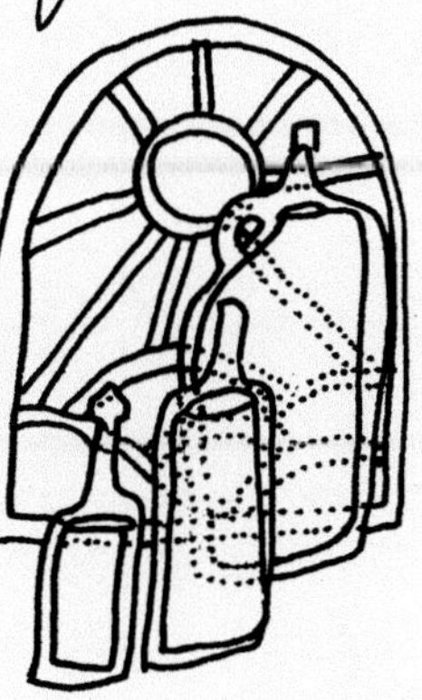

camera oscura: make a room totally dark except for a hole a few inches in diameter facing the rising or setting sun. this hole will project onto the opposite wall whatever is outside the hole in the wall, upside down. If you reflect the image onto the floor with a mirror, the image will be right side up when your back is to the hole in the wall:

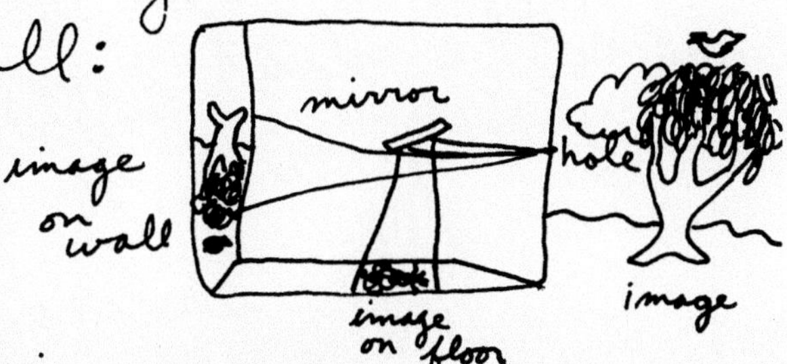

Solar attenuators (light volume control)

latticed arbor, trees, hair, eyelashes, window shades, clouds, houses, hats:

straw hats break up the sun into tiny points of light. a cloth bag of white, blue or purple cloth is good for meditating on sunlight (over your whole head). old-fashioned sunbonnets make pleasant shade.

a sun temple

is an environmental sculpture given to the interplay of light & shadow. it might include a sunstrobe ceiling like the one described a few pages back, plus warm baths, each with a different color glass skylight (blue for serenity, green for healing, violet for ecstasy). also latticed passageways, stained glass windows, solar & wind-operated musical instruments, mirror balls, prisms, a greenhouse, a sundial, domes, and the solar light shows on the last pages would enhance it. and, of course, the proximity of a sacred grove to complete it.

BEING OF THE SOUND

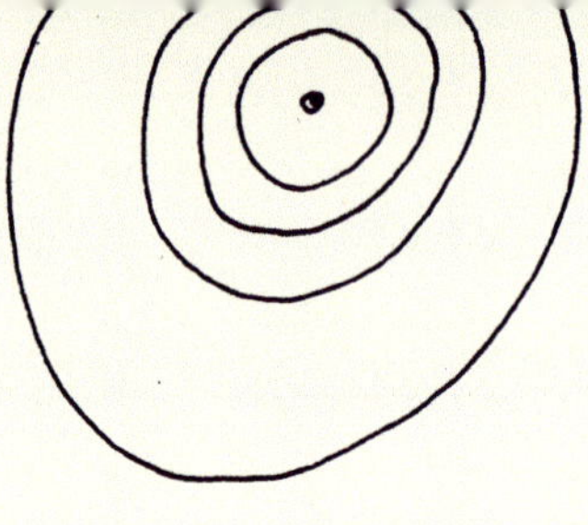

anything that exists can be perceived as a source of vibrations. sound (existing in time) and color (existing in space) vibrate in spherical shapes from their sources.

the waves made by leaves dropping onto a still pond are like your own vibrations going out into the universe. sound vibrations are called tones. the tone A is 440 vibrations per second. Benignity & malice are perceptible as "good vibes" & "bad vibes". To feel these vibrations in populous places is more difficult than in quiet places with fewer sources.

white noise

just as white light is all colors occurring simultaneously, so white noise is all pitches of sound occurring simultaneously. The wind speaks in white noise; ocean surf, the voices of trees, meadows and fire speak in white noise. the earth rumbles deeply in white noise. the colors of white noise are the vowels ah, ay, ee, oh, ooo, eye, eh. Vowel-tinted white noise is a good way to speak to earth, water, air & fire. our name for the sun is IAOUEH; the earth is OH-AH; the wind, A-OH. (whisper). water speaks in many voices, dripping, trickling, pouring, swirling, roaring, babbling, bubbling, gurgling, falling, lapping, splashing, crashing, boiling, cracking, tinkling.

at night the static is freer of other sources such as nearby electrical appliances or automobile engines

drones

a drone is a repeated tone or tones, or white noise. drones may be used to back up melodies, especially repetitive mantras or music played in modes (more about modes later). engines drone on tones, the harmonics of which you can hear as they approach & recede. when riding in cars, chant with the drone of the engine. everyone can sing a different song at the same time with a drone because they are all in the same key. instruments may be tuned to play drone backgrounds for chanting or you may tune your voice to the natural drones of crickets, flowing water, your footsteps or the wind. groups of voices and instruments can make improvisational music around a drone, taking turns droning and making melodies. birds like to be tuned to and will sing along. your breathing & heartbeat are drones. listen for drones in your environment.

harmonics

(also known as overtones. any guitarist can play some for you.)

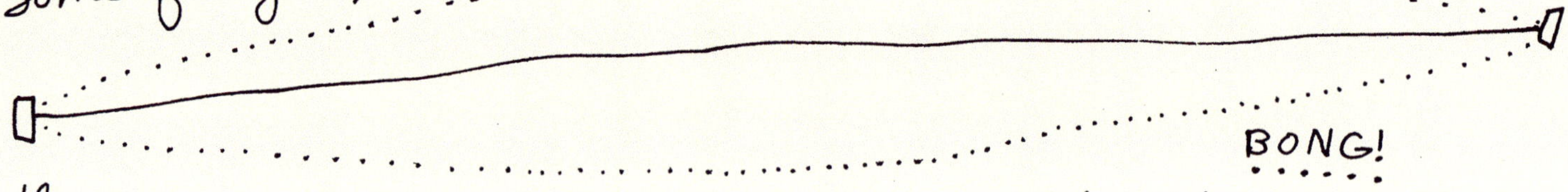

this is a string vibrating as a whole. It makes a sound when plucked or bowed which is its tone. (middle C for example.) If you lightly touch the string exactly ½ way from one end to the other and then pluck or bow it, the tone will be an octave higher (vibrating twice as fast). this is the first overtone of the string, which, in this example, would be C above middle C. Here is the first seven of the overtone series on this string:

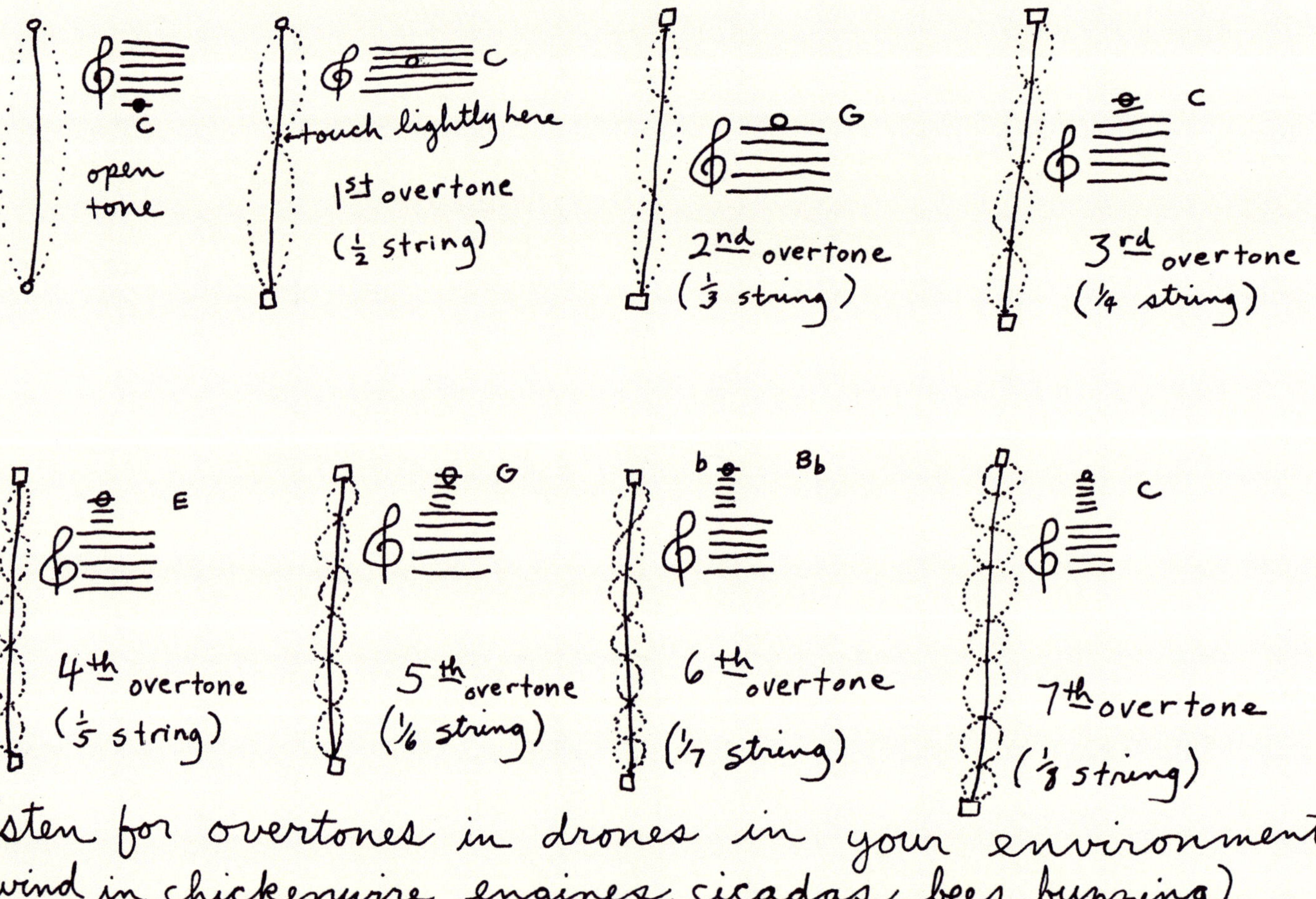

listen for overtones in drones in your environment (wind in chickenwire, engines, cicadas, bees buzzing).

as you go up the overtone series the interval between the tones decreases, so that the 14th overtone is approximately a quarter of a tone above the 13th overtone.

When you play a tone, all of the overtones are sounding simultaneously, though you have to listen intently to hear them. a low piano string has hundreds of overtones.

hum one tone. draw your lips into a little O and gradually open your lips as far as they'll go (as if saying WAHHH). be sure to let your palate resonate. You can make the individual overtones sound above the basic tone you are humming (especially the 4th, 5th & 6th harmonics).

the solar system as a vibrating string:

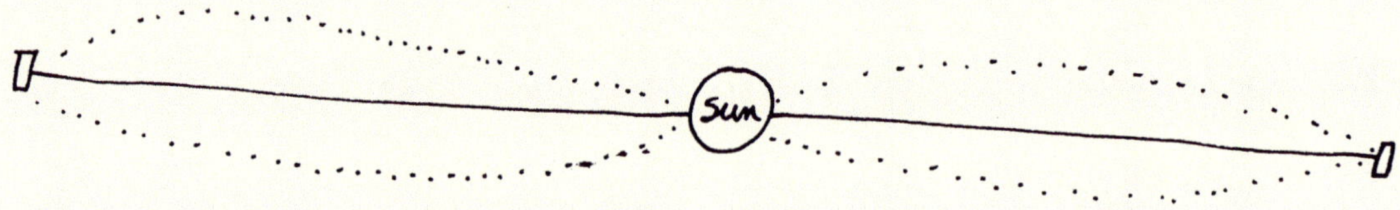

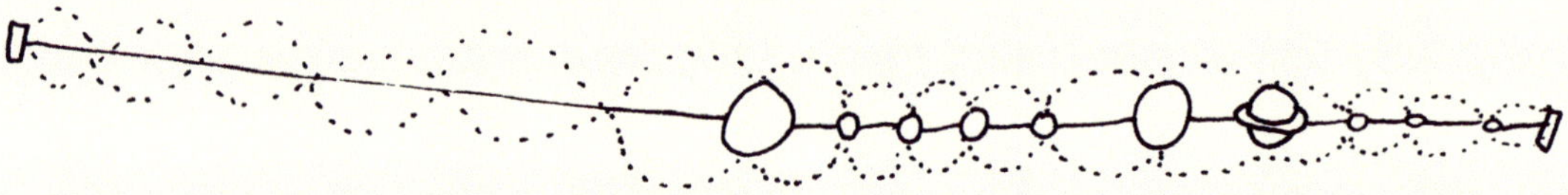

planets as harmonic points on the string.

the spine as a vibrating string:

the nerve centers (chakras) are like the overtone series. If you meditate upon them you can learn their individual tones and hum their tones to stimulate them to health.

tuning by beats

when you are tuning strings to the same note, listen to the sound they make in combination. two strings that are a whole tone apart, for example C and D, make a pulsation of loud to soft (wah wah wah), with many beats (pulsations) per second:

(this line is one second long)

if the tones are just a fraction out of tune:

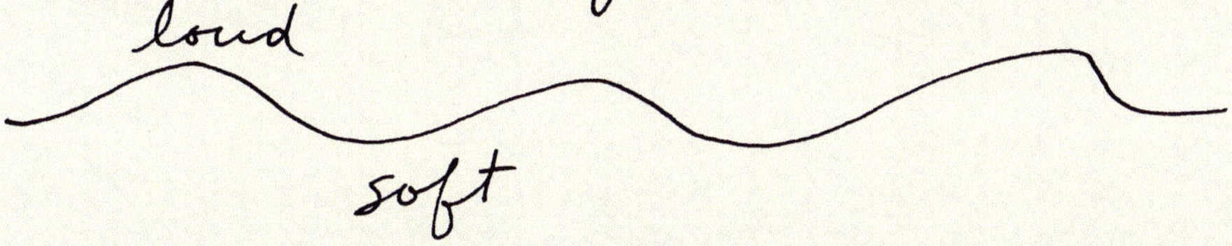

if the tones are in unison:

loud

soft (there is no pulsation at all)

Ramón's mouth moves slightly as he tunes, perhaps resonating at the correct tone to aid hearing whether the string is sharp or flat (too high or too low). a guitar pitchpipe or a piano can help you find the tones. be careful not to tune strings too high or they will break. if a string won't push down more than half way, tune *down* to the appropriate tone to keep moderate tension.

autoharps

° ° ° ° ° ° ° ° ° ° °

for drone tunings, an autoharp with the machines removed has no equal, but it takes patience to tune the 36 strings. you'll need a zither tuning hammer to turn the pegs. the wire gauges vary from harp to harp, so select similar gauge strings to tune in unisons. if a string can be pushed against the soundboard, it will take a tuning up. if it can be pushed a little it is at optimum tension. If it won't push at all, tune down. If you're tuning up & the pitch isn't changing much, STOP: you're at maximum tension & should tune down. old strings especially get fragile & break easily. autoharps (or zither or santur) can be played with your fingers, with wooden mallets, with flat or finger picks, with a stick (gently stroke with the tips), a feather; good playing sticks may be found in a grove on the ground. also they can be played by babies, kids, anybody (after they are tuned properly) and are practically indestructible. Have four, pass out 3 to whoever will play, and keep tuning one, circulating it & tuning again whichever is getting out of tune (temperature affects the strings a lot). the sound is celestial.

 open tuning for autoharp on the next page:

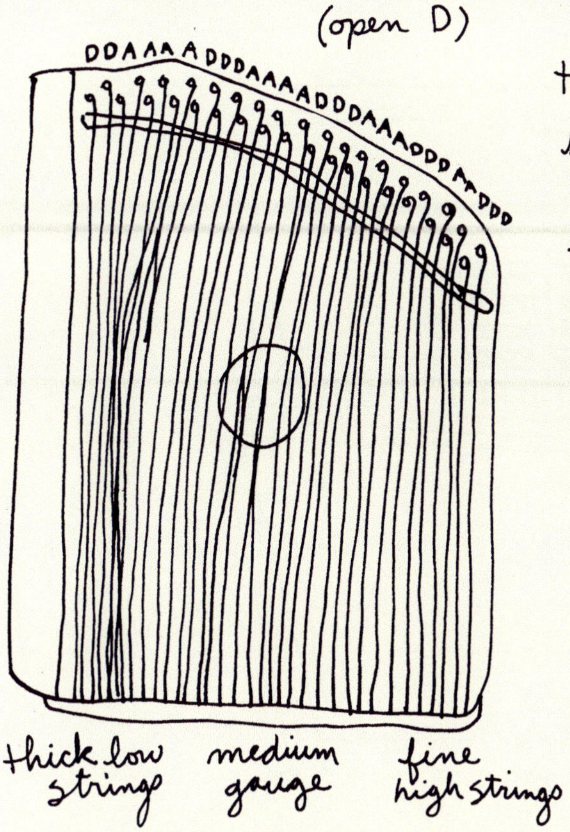

this tuning will go with any of the chants in this book. for variation, add some E's and some B's or try adding to these four tones some C's & some G's but these additions will limit the number of modes you can play with. you can tune in all the notes of a mode in order of their melodic complexity (1st, 5th, 7th, 4th, 2nd notes of modal scale - more about this farther on). the more pitches you add, the more you should pluck melodies instead of stroking out drones.

guitar, dulcimer & violin

these shouldn't be tuned more than a tone above or a tone and a half below their concert tunings: (all lowest to highest in pitch)

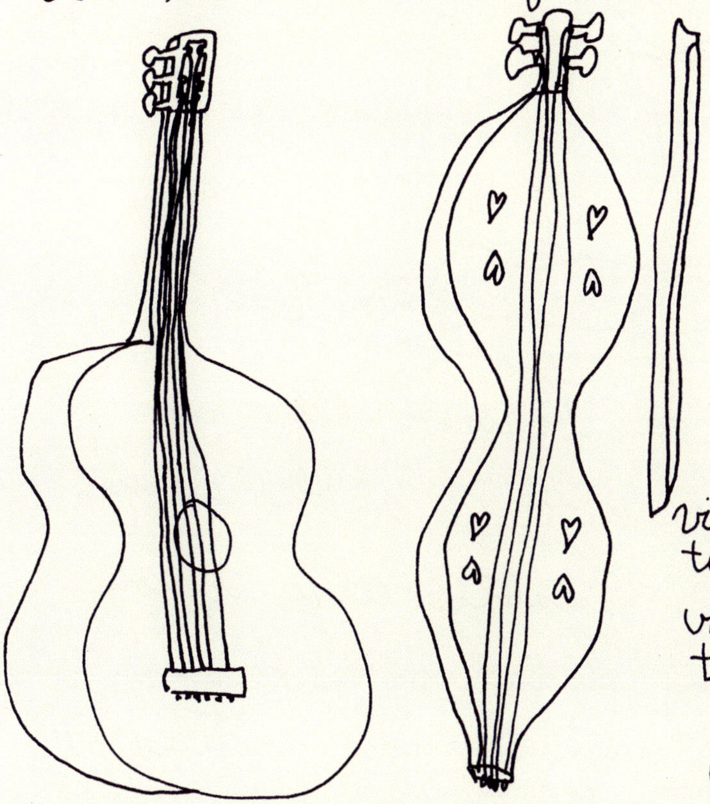

guitar concert tuning:

EADGBE

guitar open tunings:

DADDAD (open D)
DADF#AD (d major)
DADFAD (d minor)
DADGBD (d modal)
DGDGBD (g major)

violin concert tuning: GDAE

violin open tuning: ADAD (open D)
GDGD (open G)
(for drone bowing)

dulcimer tunings:

DADD (open D)
DGDD (open G)
DAC#C# (D 7th modal)

note: the modal tunings are for melodies rather than droning.

the insides of old pianos can be used like an autoharp. several people can play it at the same time. lay it on its side on a table. also if you wedge down the sustaining pedal on a piano, the strings will reverberate with sounds around it.

drone orchestra

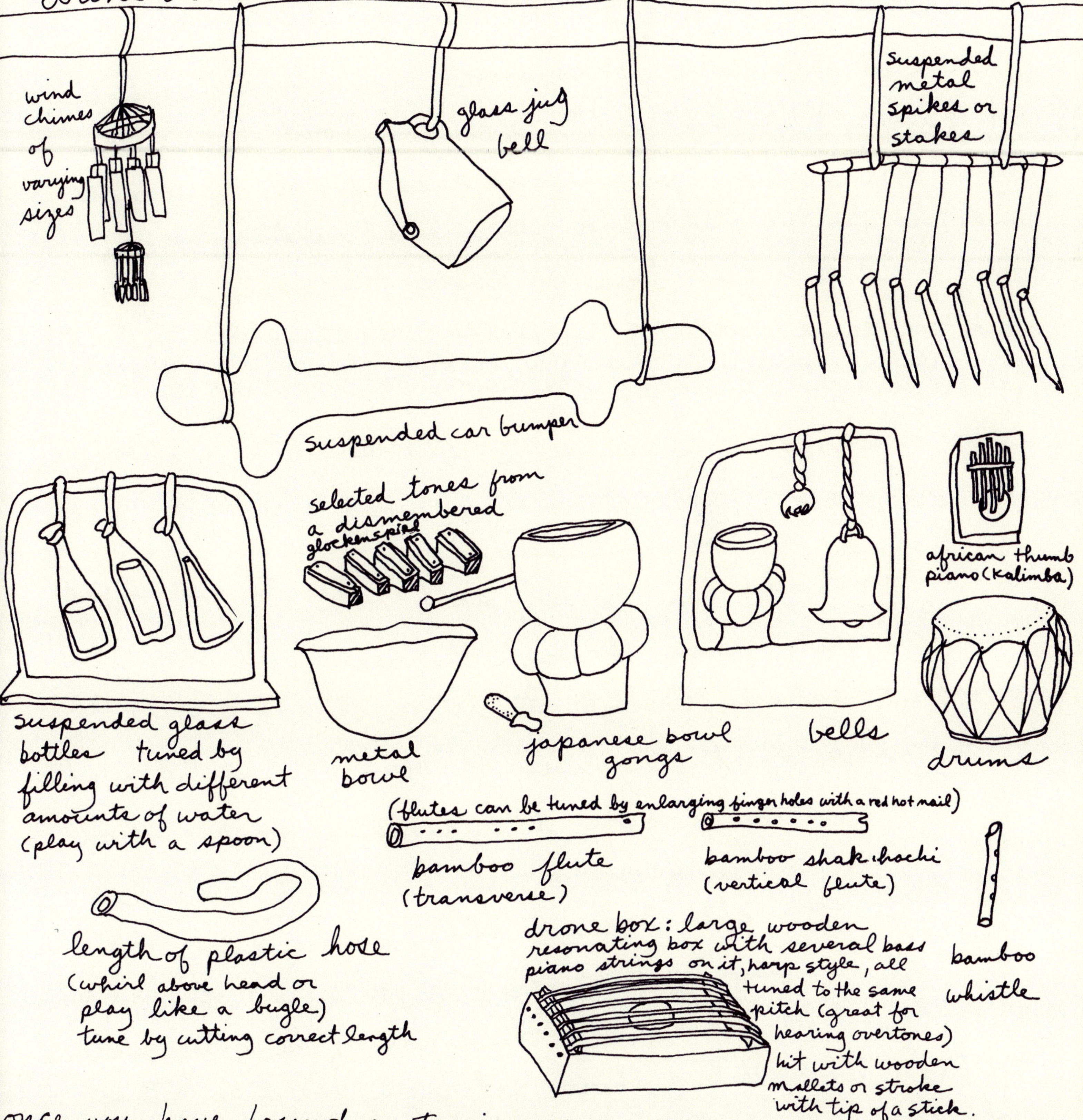

once you have found a tuning you like to play in, you can collect or make things which make tones in harmony with your tuning (or mode). the drone orchestra can be used by children, "untrained" musicians, "non" musicians, anyone, to make improvisational music or to accompany the chanting of mantras.

"homers" or dance whistles

junk a broken-down accordion by removing the whole set of sharp & flat reeds (the black keyboard notes). the reeds are set in pairs, one reed for the inbreath and one for the outbreath, in metal rectangles of varying sizes. cut bamboo tubes the same length and the same width of the metal rectangles. mount the rectangles inside whichever bamboo tube fits each one best. block with masking tape or pieces of wood the upper half

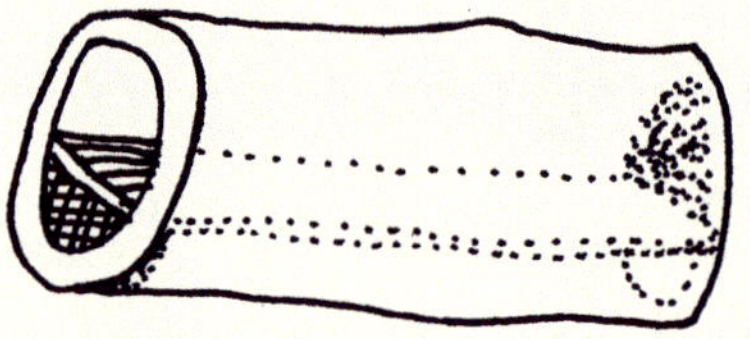

of one end and the lower half of the other end so that air must pass through the reeds whether you inhale or exhale through it. tie a string around each one so that they may be worn around the neck. pass out all the homers to a group of people. you can dance holding one in your mouth guitar & autoharp can play behind them, droning on the same notes (C# D# F# G# A#). If you want to make another set of homers, use the reeds of the white notes CDFGA. and still a third set from the E's & B's. color key your sets with ribbon or tape so you don't mix them up, and don't pass more than one set out at a time! large accordions have many different sets of reeds & can supply many sets of homers which can also be used in drone orchestras.

oat straw flutes

in california, the wild oats that grow by the roadside, when mature & dry, provide a tube about this circumference o (similar to a drinking straw). cut a notch below the closed end of a segment with a sharp razor:  gently coax up the reed you have cut until it rides like this:

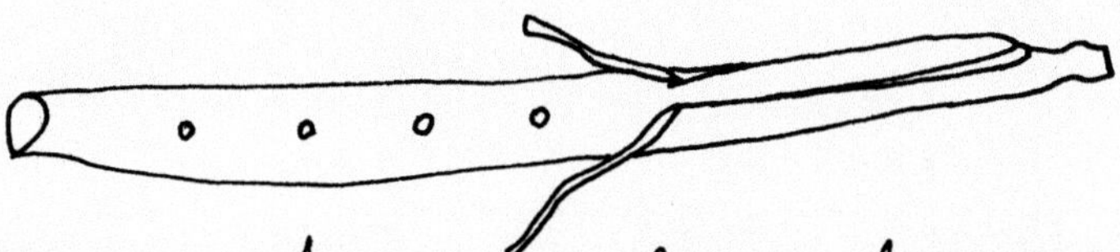

(you can place a thread at the end of the notch to keep it open.)

if you make a whole bouquet of flutes, you'll have over half of them playable. place the flute inside your mouth all the way to the end of the notch. trying to keep the reed dry, blow as if blowing out a candle. if no air & sound comes out, try coaxing the reed to ride higher. blow again. if air starts through the reed and it clicks shut, it is not strong enough. it takes patience. tune it higher by shortening the tube. burn finger holes in them with a red hot nail (held with a pair of pliers). with practice you can play two or more at a time, a drone reed out of each side of the mouth and a chanter (melody player with finger holes) in the center. also these flutes may be used as reeds for a homemade bagpipe or a bamboo root oboe. a wide tube of bamboo makes a quiver for sets of flutes.

homemade bagpipes

close the open end of a plastic bag with masking tape. tape a piece of cardboard or leather to the end of a plastic tube to make a one-way valve (hinges open when you blow through the tube but can't open the other way.) insert tube into a hole in one corner of the plastic bag, valve end first. seal with masking tape. choose 3 oat straw flutes of similar size & strength - trim to tune them to the same tone (or one drone may be a fifth higher than the other). make 3 holes in the other 3 corners and tape in 3 bamboo tubes that will hold the oat reeds in place (the oat reeds are replaceable). insert reeds into bamboo holders, notched end first. one reed has finger holes to make melodies. hold bag under one arm with the drone reeds in back, the tube in your mouth and chanter in your hands. blow into the air reservoir and squeeze it gently with your arm. the reeds should all respond equally to a push of the bellows. wonderful exercise for the lungs, but somewhat difficult to play.

bamboo root oboe

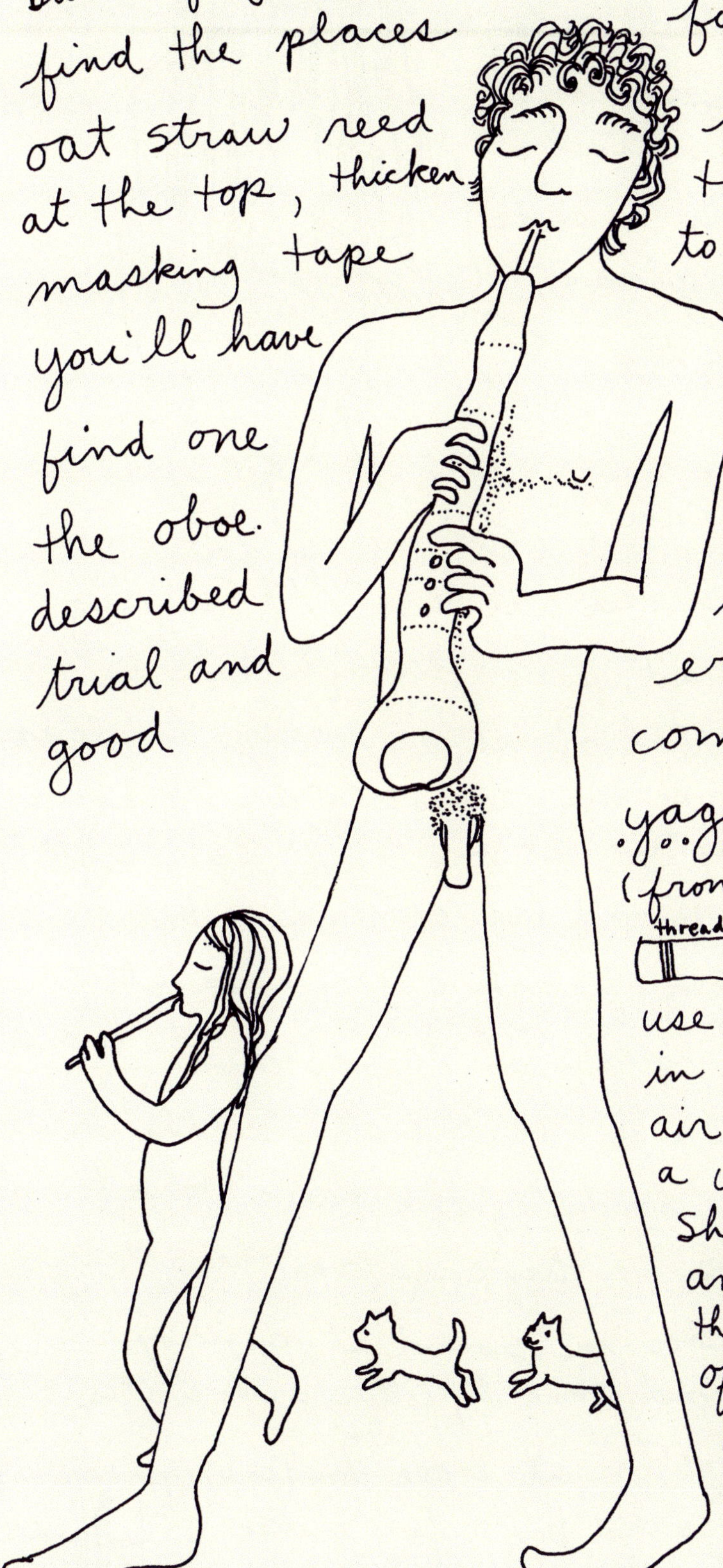

hollow out a bamboo root (well dried) by burning through the sections with a red-hot metal rod. then burn finger holes along one side. Use your fingers to find the places for the holes. to make the oat straw reed fit snugly into the hole at the top, thicken the reed stem with masking tape to an appropriate diameter. you'll have to try many reeds to find one that interacts well with the oboe. the interaction can be described mathematically, but trial and error will provide a good combination.

yagé ceremonial flute

(from the sibondoy tribe of colombia)

use bamboo 6½ inches long & 3/8 inch in diameter, a hollow tube. burn air hole 3/4 inch from one end. make a wooden plug to fit inside that end. Shave a little off one side of the plug and plug in with shaved side under the air hole. carve away underside of bamboo & plug to form mouthpiece. burn finger holes, one 1 3/4 inches from end and one 2 3/4 inches from end. bind with thread to keep the bamboo from splitting.

wind chimes

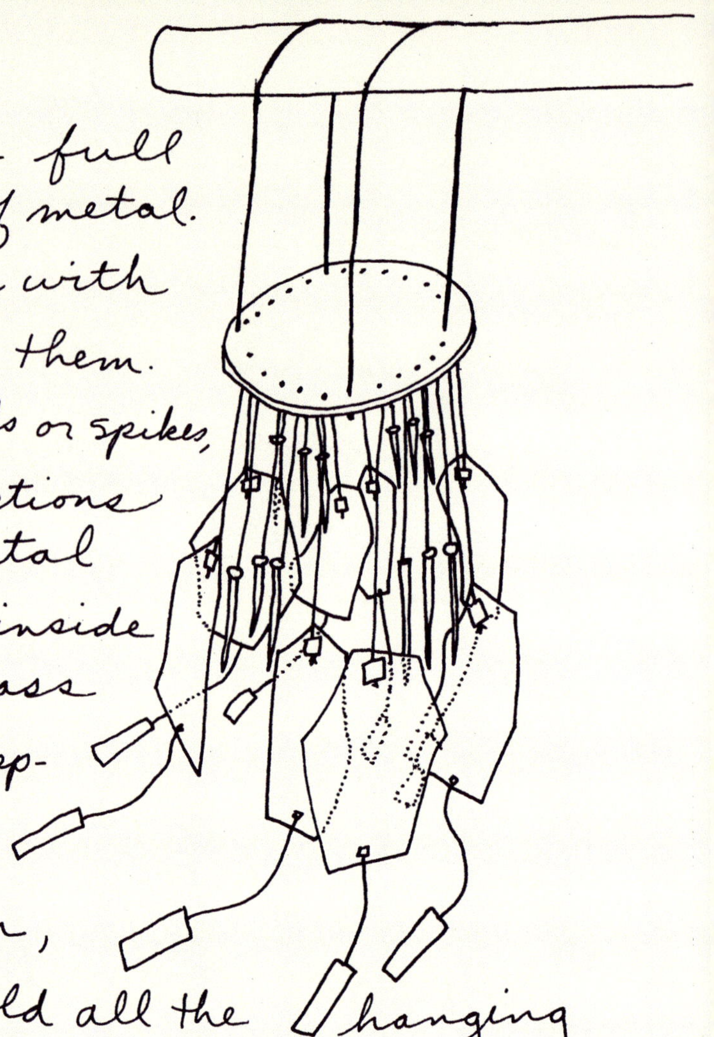

automobile junk yards are full of sweetly resonating pieces of metal. hang them by a string and tap with another piece of metal to test them. some favorites: long thick nails or spikes, bicycle gear plates, ¼" to 1" sections of aluminum pipe, heavy metal circles, telephone bells (from inside old phones), large pieces of glass from broken windows (make deep-voiced chimes).

the metal top from a coffee can, punctured with nails, can hold all the hanging things (knot a string at one end, pass through hole in coffee can top and tie to metal object or epoxy to glass with a piece of paper or leather over the string & let dry 24 hours). pass four strong cords (knotted at one end) through the bottom of the coffee can top and tie around a beam. glue or tie strings to the bottom of each object and glue to the end of each string a paper sail to catch the wind.

wind-driven drones. make a windmill with brushes or wooden beads on strings at the ends of the sails that play an open-tuned autoharp as they pass over the strings. place bottles or tubes of bamboo in sand at appropriate angles to be played by the wind. tune by filling with water. tune strings on drone box together at loose tension. place a wooden board parallel to the sound box a few inches from strings and place so wind goes between strings & board.

water drones: listen for the tone of a waterfall of a a small cascade. carefully move the rocks at the base of the other waterfalls until their tones are the first & second overtones of the deepest sounding waterfall.

vessels of metal, glass or wood can be placed under falling water (off a roof) to make melodies with water.

earth drones. a whole pasture full of cows with harmoniously related bells around their necks.

chanting. when not absorbed in the sounds of nature you can join the birds, water & trees in lifting your voice in praise of your maker. The difference between a chant & a song is that a chant is repeated many times and can be accompanied with drone music. (also see "mantras")

if you're chanting indoors, sing up & down the scale, listening to the sound. on a certain tone the room will "boom" - you will have found its resonance. On that tone chanting & droning will have the full deep echo of the room singing along.

for outdoor chanting we have evolved a set of modes & keys to sing & play in which we feel are harmonious with the hours of the day and the phases of the moon. first we choose the key by the phase the moon is in. let's say it's a full moon and the time is 6AM. the drone would be in C and the mode would be mode 7, which is a major scale with a sharp 4th and a flatted 7th. sing the scale & play on your instrument a few times to learn it. then you can improvise a song from it or use the scale (by numbers) to change the melodies of the chants in this book. if the scales are too hard, chant on the tone of the moon, or speak it.

music chart for ceremonies & chanting

(with colors for environment)

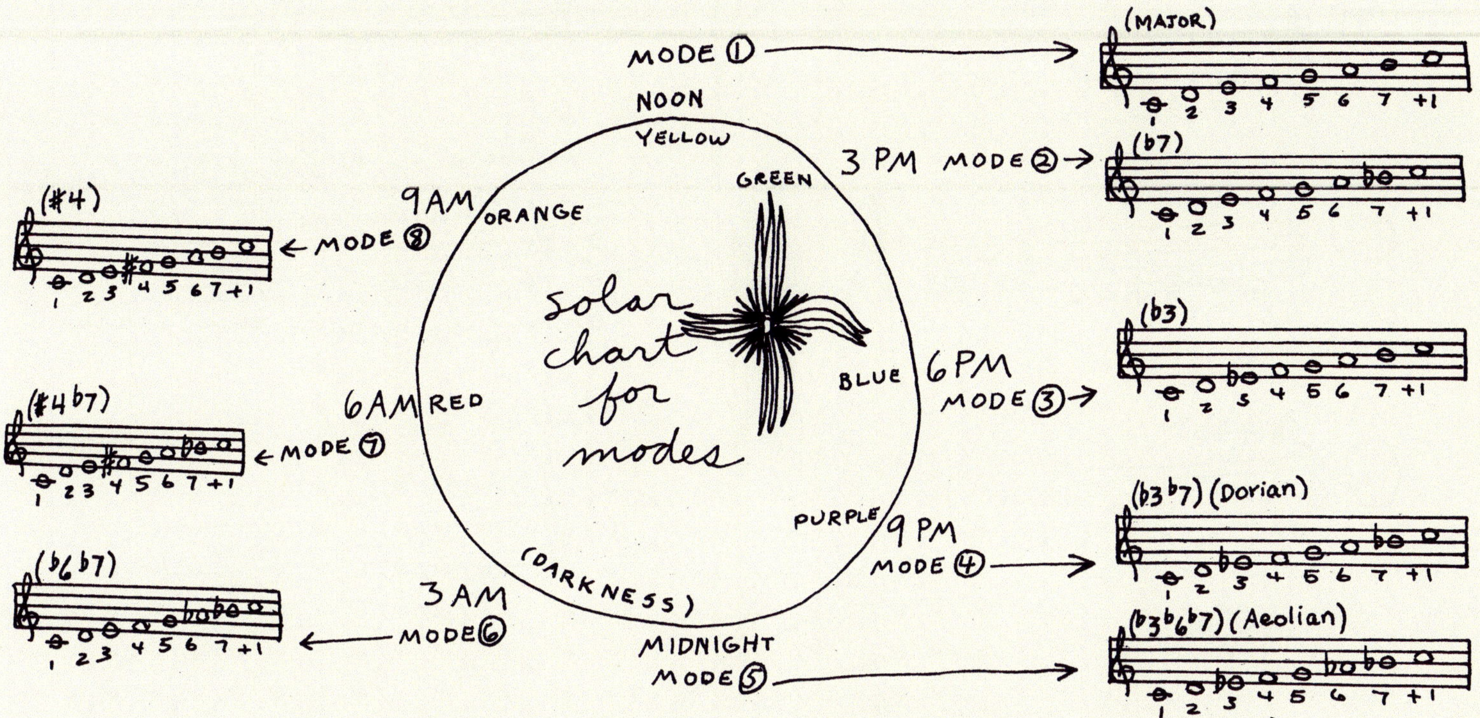

(you go to bed flat and you wake up sharp)

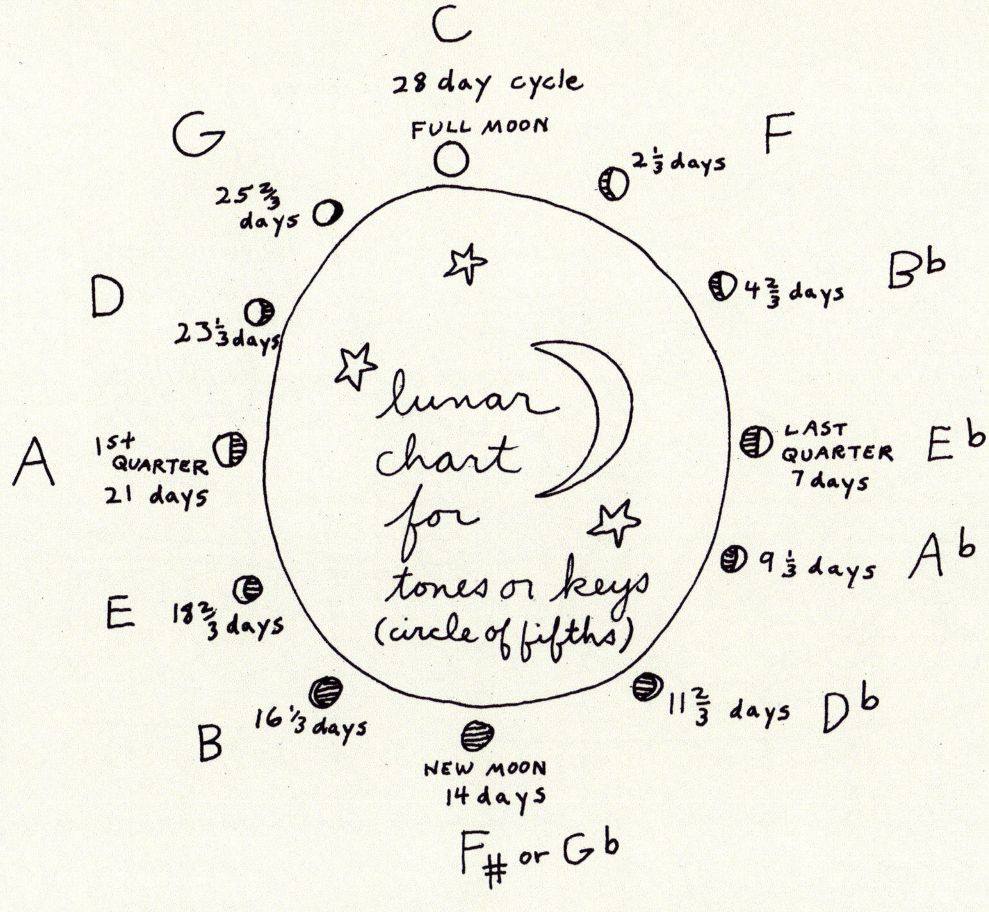

cloudy day modes these are more difficult & more exotic and complex:

(9) (♭2♭3♭6♭7) (Phrygian)

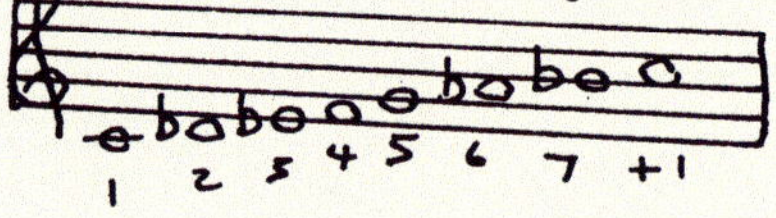

(10) (♭2♭3♭7)

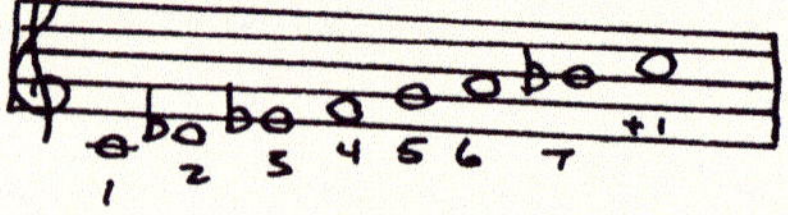

(11) (♯4♯5)

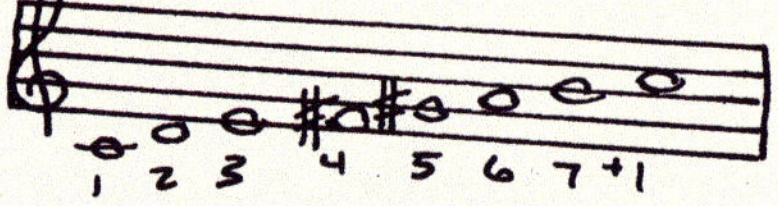

(12) (♭3♭5♭6♭7)

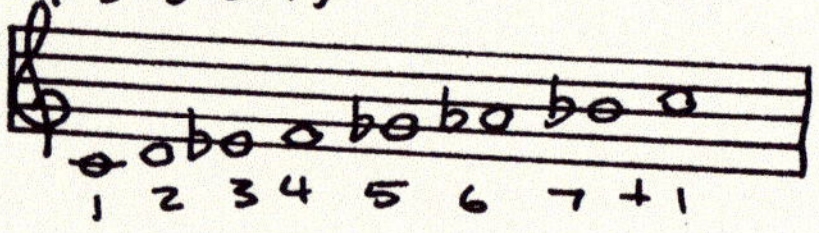

7 others may be derived by starting on

(13) (♭3♭6) (harmonic minor)

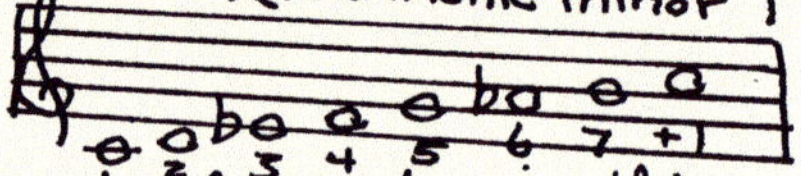

any of the notes in this scale.

this is one of Ramón's favorites:

(14) (♭2♭3♯4♭6)

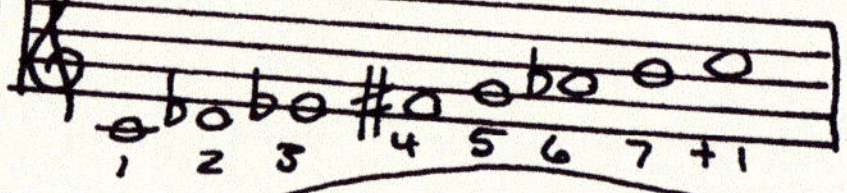

the chants in this book have numbers below the notes. if you read them by the numbers, you can substitute any of these modes for the ones the chants are written in.

twelve hands across the sky

every earth day
deserves a
celebration
hooray!
it's
today!

how to tell time by the sun

with your hands: stand with one arm outstretched toward the place where the sun came up. make the thumb side of your hand appear to rest on the horizon and the little finger side parallel to the horizon (palm away from your face). each distance across your hand above the horizon is one hour. there are 12 hands across the sky during the equinoxes, from 6 A.M. to 6 P.M.

9 AM

sundial: find a round stump cut flat or canted to the south. Set a stick straight up in the center & mark hours by observation (12 noon the shadow points due north). as the seasons change, the shadows change, so you'll have to mark new ones. After a year it will be complete.

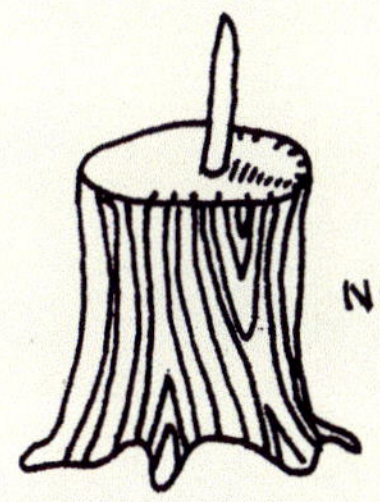
NOON

observatory: in the center of a large flat meadow, preferably a hilltop, place a stone to sit on as the fixed observation point. (you may place a sundial to read from this point as well). mark the solstice & equinox sunrises and sunsets with large stones or logs at the edges of the field (these places are also the rising & setting points of the band of zodiac stars at night). the observatory can be used as a ceremonial grounds for solstice, equinox & eclipse celebrations and for full moon meditations (see holy days chapter)

a whole day with the Sun

4AM to 6AM (violet-red)

when you awaken, observe your sleep breathing (its pace, where the sounds come from in your body, what parts you move as you breathe) so that you can copy it thoughout your day. chant the AUM to remember your dreams. Observe the perfect stillness of nature, the return of colors, the return of your lover's face. the moon in its last quarter and the morningstar rise before dawn. good time for meditation.

6AM to 8AM (picture next page) (orange)

all of nature greets the sun; songs of birds fill the air; roosters crow; sunflowers turn their heads; dew begins to rise. take your morning shit & analyze your health. Bathe or swim or just splash your face and back with cold water. do hatha yoga or other stretching dances. make bed, open windows, arrange house & altars. sing morning songs. time for milking & feeding of animals. breakfast with fruit and herbal tea. gather herbs when in full flower.

In The Morning

ABL '71

8AM to 10AM and 10AM to noon (yellow)

the morning work hours: gardening, playing with children, building, making beautiful things, dancing, hiking, walking meditation. survival work or thankless tasks done without thought of reward. observing nature.

seven work-hour ceremonies:

spend the time alone, silent, listening, not thinking

spend the time silently with someone you love (develop telepathy)

feel all muscles, bones, nerves, organs, skin moving in a relaxed way.

make an anonymous gift or good deed.

whistle with the birds; learn their language.

feel the sun as lover, friend, brother, father.

rise above your thoughts & observe your self.

. . . .

when gathering food, be sure to thank the plant or animal who is giving of itself, give your care in return. make your creative work a gift to the universe.

noon to 2 P.M. (yellow-green)

noon meal: nuts, seeds, sprouts, fruits, greens, dried fruits, goat cheese, fruit or vegetable juices, yogurt or whatever suits you best (those are our favorites). after cleaning up eating place & vessels, hang a hammock in a piece of perfect shade and take a nap. observe all the other animals at this hour, especially dogs, cats, cows, goats, horses, lizards, snakes, butterflies, birds, fish and bees. listen to the air, watch the clouds or the flow of water. relax all over. listen to drones in your environment and tune to them. sing the noon chant.

2 PM to 4 PM (green)

afternoon work hours are like the morning work hours, though you may have changed projects. in the planting of flowers, washing of clothes, fixing of broken things, or whatever you are doing, feel how the sunlight is helping you. if you are not working, lie in a grove and watch the sunlight come through the leaves or make music with children.

4PM to 6PM (blue)

the winding up of the day's work & the preparation of the evening meal. water the garden; milk and feed the animals; put away your tools and toys. listen to the sounds of water as it flows to the sea. Salads, soups, tea on the table, hold hands in a circle around the table and chant the AUM or the solar grace:

feed the first bite to someone else. on receiving this bite, chew it 100 times. we like to eat cloves of raw garlic with our evening salad to encourage health in the digestive tract. take your time & enjoy every bite. Clean up eating place and vessels. listen to the evening songs of the birds in the treetops.

6PM to 8PM (indigo)

watch the sunset & chant the sunset chant:

For Mode ④ sing as follows: 5 b7 +1 b7 5555 5, 5 b7 +1 b7 etc.

aum & chant to drones tuned to the crickets. listen to the night songs. meditate on the evening star, the waning of the light, the revealing of the constellations, the rising of the full moon.

night time (violet-black)

meditate on the stars being your brothers and sisters (be the sun). hug someone & feel the sun shining inside their body. When making love, first remain joined together without moving for as long as you can, looking into each other's eyes & synchronizing your breathing. as you go to sleep, try to remain conscious of what happens to your body. the hours before midnight give the most refreshing sleep.

each day is an opportunity
to truly help mankind
in your potential
to relaxation & bliss.
your peaceful vibrations
affect everything & everyone
around you.

holy days

the lunar cycle

the full moon a time so full of high energies and gatherings of water in the environment & in the body that it is often difficult to sleep. an all-night meditation may be occasioned by the full moon. we have discovered some points on which to focus which are harmonious with the signs of the moon; one is given in each monthly celebration. the highest tides of the month occur in the full moon.

the 3rd quarter gradual decrease in energy, a good time for unwinding from busy days & nights.

the new moon the lowest tides of the moon's cycle, a time of serene emptiness and silence.

the 1st quarter the regathering of energies. time to start projects and do most intense work. hope is renewed and excitement grows.

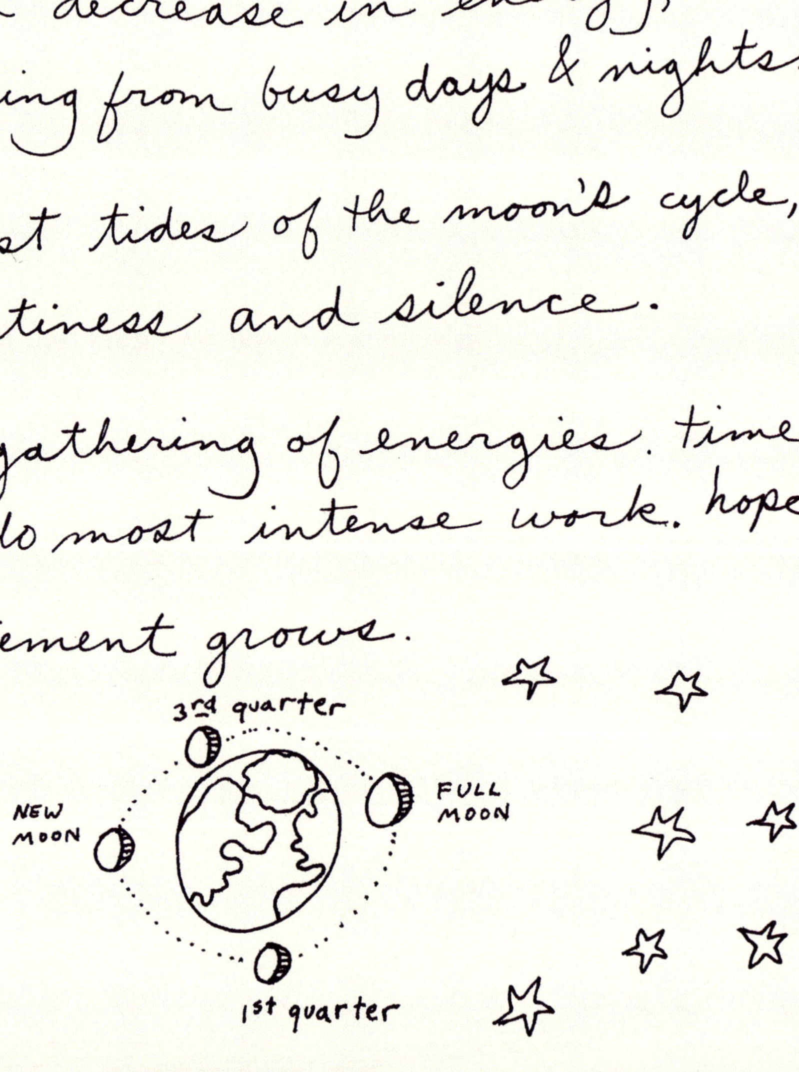

chants for the phases of the moon

tone (C)
mode (2)

Full Moon Rising

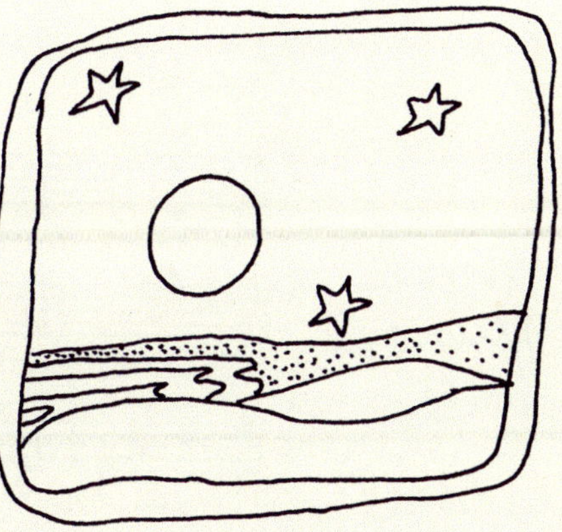

tone (E♭)
mode (4)

Half Moon (3rd Quarter)

New Moon

tone (F♯)
mode (8)

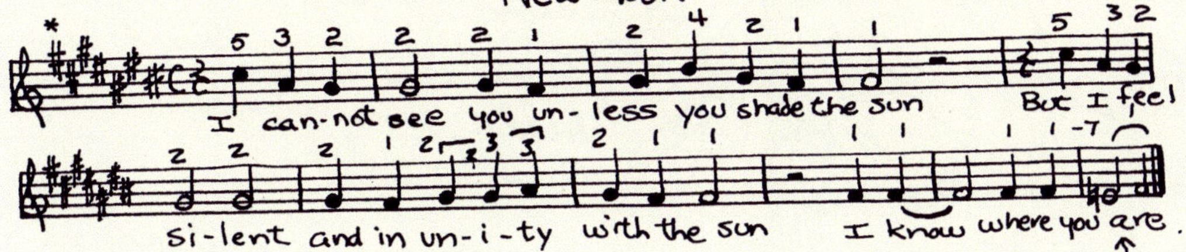

* If too many ♯'s, ignore and sing in F with no B♭ and with an E♭ here.

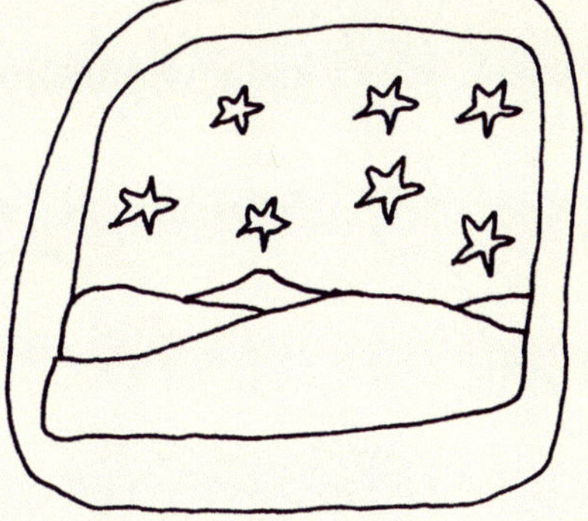

tone (A)
mode (2)

Half Moon (1st Quarter)

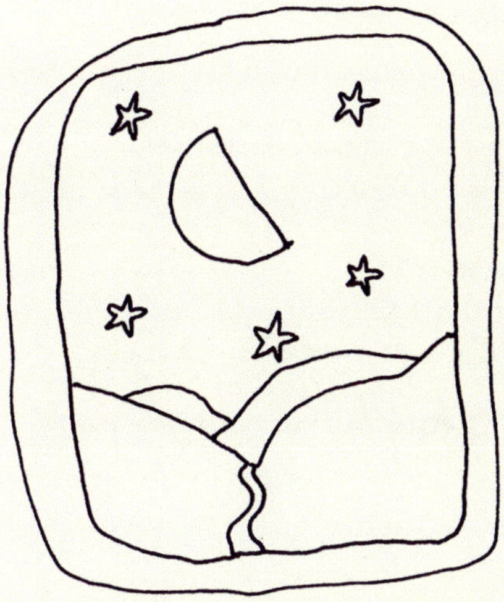

the month of aries ♈ . . march 21 to april 19

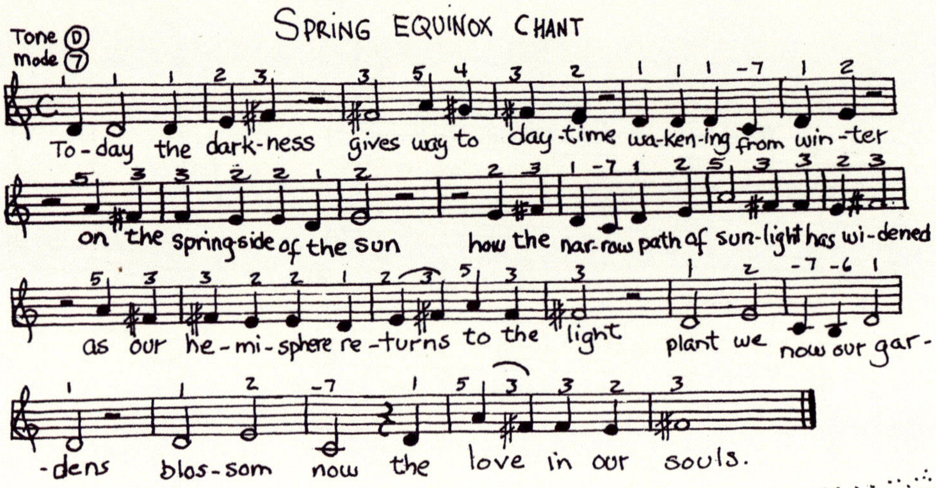

vernal equinox is the festival of air

the vernal equinox celebration: make windchimes, kites, wind-powered drones or a wind machine and assemble on a hilltop or a beach. sing spring equinox song & other drone music accompanied by the wind. AUM over seeds, seedlings and baby animals. make egg tree & egg worlds, flags & banners.

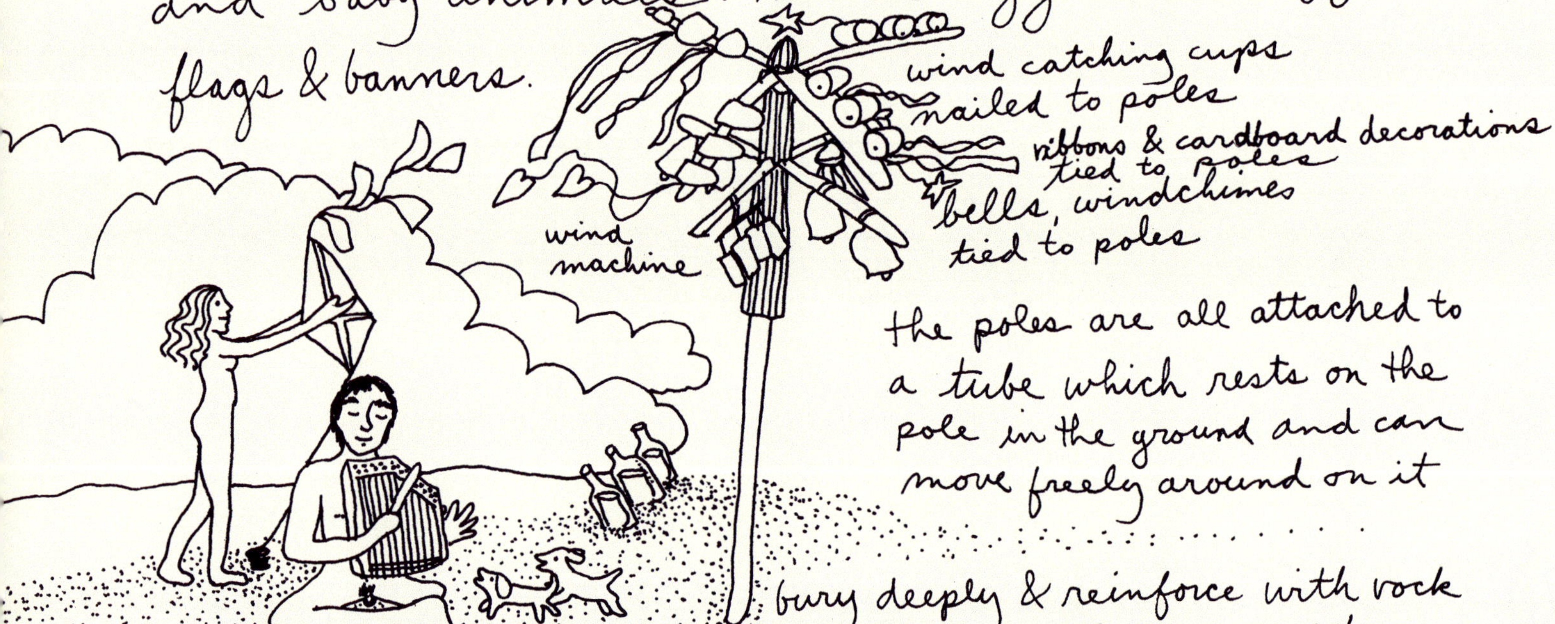

the poles are all attached to a tube which rests on the pole in the ground and can move freely around on it

bury deeply & reinforce with rock or tie to stakes on 4 sides. it is very top heavy.

full moon in libra (full moon is always in the opposite sign of the sign the sun is in) – meditate upon the sky & the upper atmosphere.

egg tree

make a small hole with a nail at both ends of an egg, and blow through one hole so the yolk & white come out the other end into a bowl. (use in a cake or omelet.) dry the shells overnight. decorate them with paint, beads, tissue paper, lace, ribbon, paper cut-outs, etc. (white glue works well). glue thread or ribbon to one end (with a bit of paper over it to secure it) and dry thoroughly. tie them to a potted tree or a fallen branch potted in rocks or hung from the ceiling. this is an indoor decoration-weather can hurt the paper. after the holy day, wrap the egg shells in tissue paper, store in a box for next year's. my (A) mother has some I made 15 years ago!

egg worlds

after blowing out & drying the egg shell, widen one hole enough to build a little scene inside it. old dentists' tools, tweezers, toothpicks can help you glue the tiny paper cut-outs in place. the scene faces the smaller hole - when the construction is complete, glue a piece of transparent tissue paper over the big hole to be the sky behind the scene and look in through the small hole, aiming the sky at a light. my (R) sister makes little forests & chapels with stained glass windows! carefully decorate the outside of the egg, too.

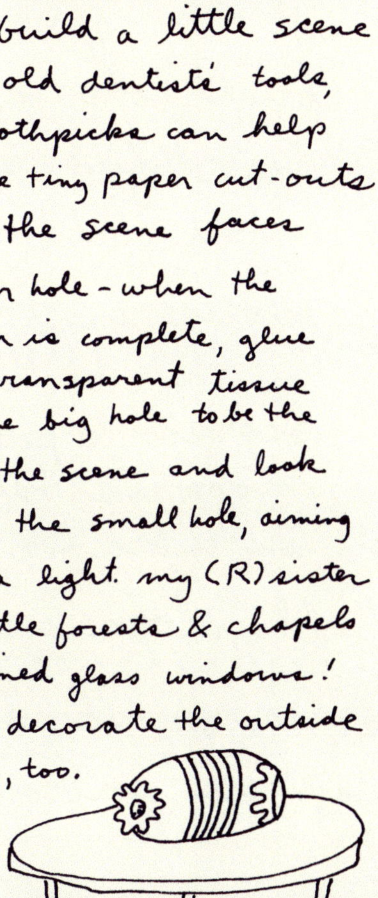

the month of taurus ♉ april 20 - may 20

apple blossom time: go out in an orchard & listen to the songs of the birds. select five of their favorite tones to make a 5 tone mode. then get 10 to 50 pieces of aluminum conduit pipe and cut them so that they chime on your 5 tones. drill two holes near one end of each chime and hang in pairs from the trees so that the wind will knock them together and make the whole blooming orchard chime with the songs of the birds.

planting a garden after the soil is prepared to receive seeds and young plants, decorate the garden fence with streamers, god's eyes of bright colored yarn lashed around sticks, a beautiful scarecrow & other gestures of good will to the plants. Plant fruiting & leafy crops in the waxing of the moon and root crops in the waning of the moon. the water signs (cancer, scorpio & pisces) are the most fertile moonsigns. moon in taurus is also good for root crops and moon in libra for the planting of flowers.

full moon in scorpio: meditate upon the arctic poles, on ice & snow & tundra, glaciers & snow-capped mountain peaks.

planting day ceremony

CHANT

(to the earth and sun)

CHANT

(to the plants)

may day

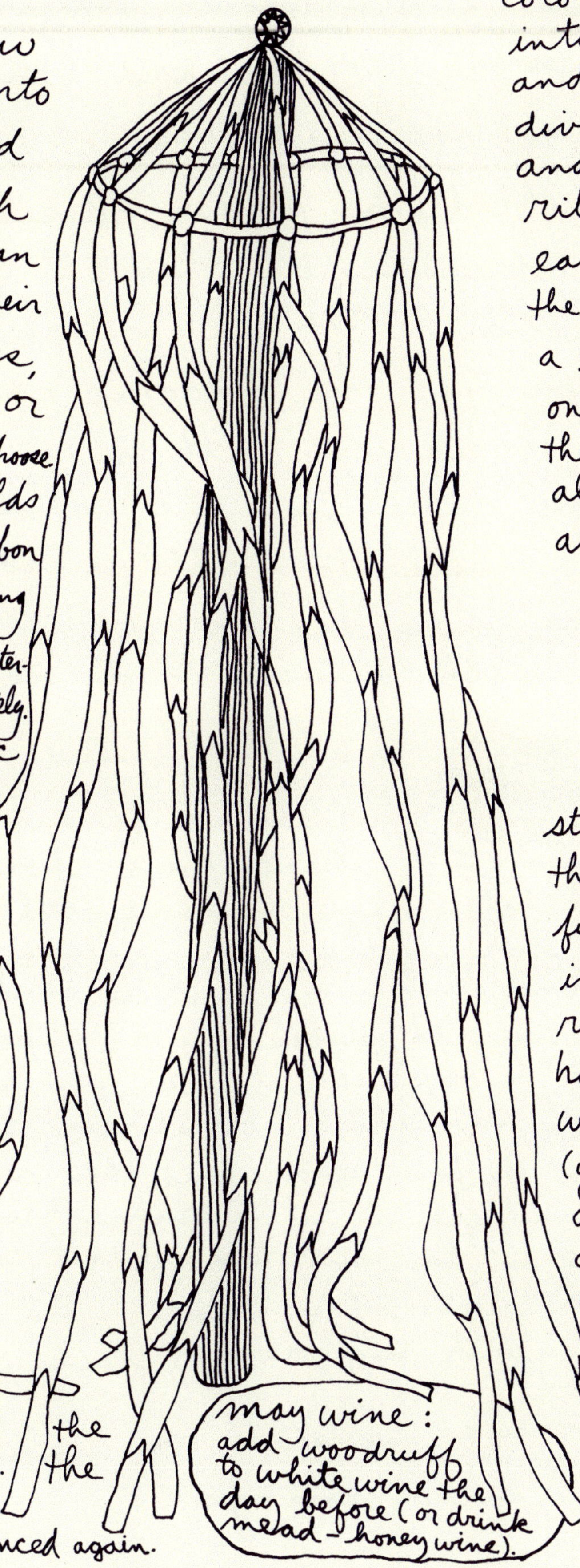

dancers: sew jingle bells onto bright colored ribbons which the dancers can tie around their wrists, ankles, arms, waist or head as they choose. each dancer holds the end of a ribbon of the pole facing clockwise & counter-clockwise alternately. When the music (drone of harps and flutes) begins the clockwise-facing dancers go outside the first person they meet, inside the next, outside the next and so on. the counter-clockwise do the same only they start by going inside the first person they meet, until the pole is wrapped. the pole may be unwrapped & danced again.

may wine: add woodruff to white wine the day before (or drink mead-honey wine).

may pole: tear up bright colored rags & old clothes into strips 2 inches wide and as long as possible. divide into piles by color and tie together to make ribbons 20 feet long, each a different color of the rainbow. tie by cutting a slot one inch long, one inch from the end of the strip on both ends of all of the strips. then attach:

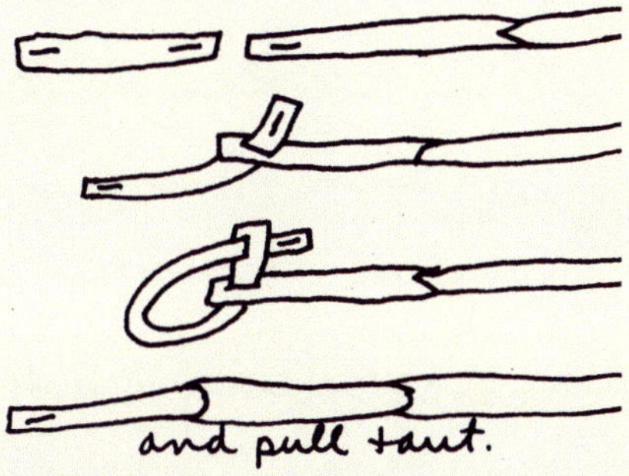

staple all the ribbons to the top of a pole at least fifteen feet long & 3 inches in diameter. then tie each ribbon once around a barrel hoop a foot or so below where they were stapled. (you can add extra streamers or a cut-out decoration on top). dig a hole (with a post hole digger) 3 feet deep in a meadow suitable for dancing. a may pole made this way can last for years.

month of gemini II may 21 - june 20

weddings: choose a good planting day & have the ceremony near a garden or orchard (the growing things give & receive blessings on these days). when everyone has assembled, the person closest to the couple imparts a blessing as they sit in two chairs. then all the men lift the bride's chair and all the women lift the groom's chair and carry them in a procession with music to the table where the cake is. the couple cuts the cake & gives a piece to each person at which time anyone may bestow good wishes upon them. then much music & dancing.

wedding cake

make 3 round rich fruitcakes of three different sizes using whole wheat flour, honey, chopped nuts, chopped dried dates, figs, raisins (roll each piece in flour so they don't stick together), tahini (sesame butter), and grated orange peel, ground cinnamon, nutmeg, cardamom & a little salt. cover a tray with pieces of wax paper and place the largest cake on top. ice it with a mixture of honey, shredded coconut, and tahini well stirred. place the second largest on top of the first and ice it, then place the smallest cake on top and ice it. cut an orange through the center and place a slice on top. cut thin wedges to decorate the sides. pull out the pieces of wax paper from under the cake. decorate the tray with tiny vases of fresh flowers.

full moon in sagittarius meditate on lightning, people's auras, the flow of energy.

an alone wedding ceremony. find a magic place in nature and make an altar with a comfortable place to sit, musical instruments & a bowl of fruit. sit together in tantric yoga meditation, make music & feed each other slices of fruit.

love prayer. oh endless bliss
stretching from lover's kiss
through the whispering wind
through the center of the sun
to the birth of galaxies
all paths meet in thee. RS '69

the month of cancer ♋ june 21 to july 22

summer solstice (the festival of water)

swimming, sunbathing, hanging of banners, hammocks, swings and awnings. drone music & AUMing in gardens, orchards & greenhouses. on a shady beach or hilltop make a picnic of early summer fruits and vegetables with much singing, dancing and lying around watching sunlight coming through the treetops or sunstrobing in the full sunlight. at high noon a great shout, blowing of conchs and beating of drums: hooray for the Sun! summer solstice chant:

full moon in capricorn: meditate upon the soil, on gardens, orchards, forest, jungle or ocean floor.

banners cut a rectangle of cloth and hem the ends over a stick of wood. tie string to both ends of the upper stick (the lower one is to weight it down). sew patches of cloth to the center to make pictures or mandalas. if you use heavy nonabsorbent fabric like canvas, you can paint on it. hang them on walls or trees on festive days.

awning... make a banner & lash two more sticks to the end sticks to hold it up. place a stone in the center and tie string around it on the outside and tie to a beam or branch. sew cloth to the sides and secure to stakes at corners using stones & string.

the month of leo ♌ july 23 to august 22

name-changing ceremony. at a tribal gathering, a person who has recently had a vision might stand up & say "to remind me of the great nameless One and to recognize my true self, I ask you to greet me as (new name)." then a drone chant of the person's new name for several minutes will help people remember it.

anointing ceremony. to be performed by four to six friends upon one person, before he or she is to be wed, or after a heroic deed, or as a group gift. massage should follow a bath or swim. Spread towels on a floor, bed or table for the recipient to lie on. use these oils:

oil of mint: center of forehead, back of neck, temples
oil of rosemary: hair & scalp (a great conditioner and doesn't leave oiliness or smell).
wheat germ oil: whole body. rub firmly, don't pinch, knead towards the heart. almond oil is also good, especially on fingers & toes.
coconut oil (if solid, warm it up): genitals, asshole.
perfumed oil (patchouli, lemon verbena, sandalwood, rose etc.) on wrists, behind ears, neck if the person prefers it to his sweat.

full moon in aquarius. meditate on the farthest reaches of outer space.

one oil anointing. wheat germ oil

foot anointing... wonderful after a long hike. wash the recipient's feet in warm water and rub them with salt while wet. Rinse in clean warm water and dry thoroughly, especially between the toes. then oil the feet and ankles.

freeform ceremonial dances

maintain a loose pelvis; you will feel pleasant sensations rush through your body when your spine undulates from the movement of your pelvis. the movement of your knees is the key. conga drums make the best dance music. earth dance: move with great effort as if through a medium of thick jello. water dance: move as if swimming, floating or skin diving. air dance: move as if flying, gliding. fire dance: radiate a great warm light from the center of your chest. unfolding to the sun: begin curled on the ground and slowly grow like a flower. group dance I: make eye contact with each person you meet. don't talk. group dance II: eyes closed, make body contact with each person you meet. couple dances: dance with one part of body (other than hands) always touching. take turns losing your balance & being steadied. simultaneously explore each other's bodies.

prasadam is a food offering to god in all beings. it can be cornmeal on a bird feeder, dried fruit offered to friends on a hike or hours working in the kitchen of a free food mission or a large commune. It can also be placed on altars in nature: a tree stump feast for local wild animals (deer like oats, raccoons like dogfood, rabbits like raw vegetable parings), or crumbs for fish & ducks in a pond, or compost in the earth for the earthworms and green growing things.

prayers are conversations with your god-self. they can be thought, spoken or written. first you name your desires — changes you'd like to see happen in yourself, in your environment. when these are listed, count your blessings and offer thanks for these. by this time if an action on your part is necessary for the fulfillment of your wishes it should be evident. little bits of advice will pop into your head from god-knows-where. listen to all your voices and then relax and let the changes happen. if it is a written prayer, sometimes it's fun to shelve it for three months and then reread it to see if your wishes had any effect or whether you outgrew them.

full moon in pisces: meditate on clouds, fog, vapor, mist; the movement of water through air.

a visit by a very special holy person deserves a feast and lots of ethereal drone music. make a necklace of fresh fragrant blooms: with a darning needle and a long piece of button thread, string the flowers together by piercing the base of each flower so that they all face away from the thread. then several smiling people can carefully place the necklace around the holy person's neck, and sing the welcoming chant.

the month of libra ♎ september 23 to october 22

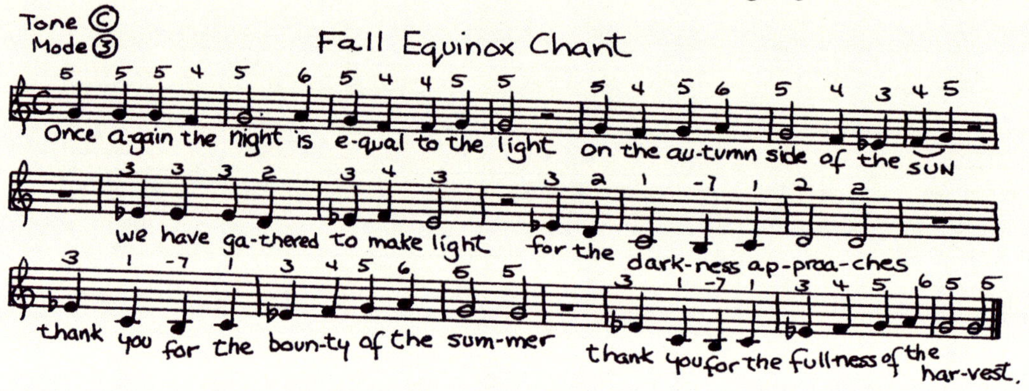

the autumnal equinox (festival of earth): rake up fallen leaves into piles and jump in them or lie on them. make a pit and light a fire in it. roast potatoes & unhusked corn, drink hot cider. AUM around garden, orchard, cellars, grain bins and canning closets. AUM over the oven while baking sunbread. at sunset go to a hilltop to sing the fall equinox song, the sunset chant & other drone music.

full moon in aries. meditate upon fires, flames, candles, bonfires, kerosene lamps.

ceremonial sun bread

make whole wheat yeast bread dou knead it and let it rise to double in a warm bowl covered with a cloth. Oil a large cast iron skillet. separate the dough into three large parts and one small part. roll the small part into a ball and place it in the center of the skillet. Roll the three large parts into ropes one to two inches thick. braid the ropes and place the braid in the skillet surrounding the ball. attach the ends of the braid so that the circle is continuous. cover with a cloth and allow to rise again; bake at 400°F (or whatever temperature recipe calls for). When your bread almost done, brush the top of with a mixture an eggyolk and one tablespoon of water. bake 10 minutes. hand-sized chunks can be torn off as the bread is passed around a circle.

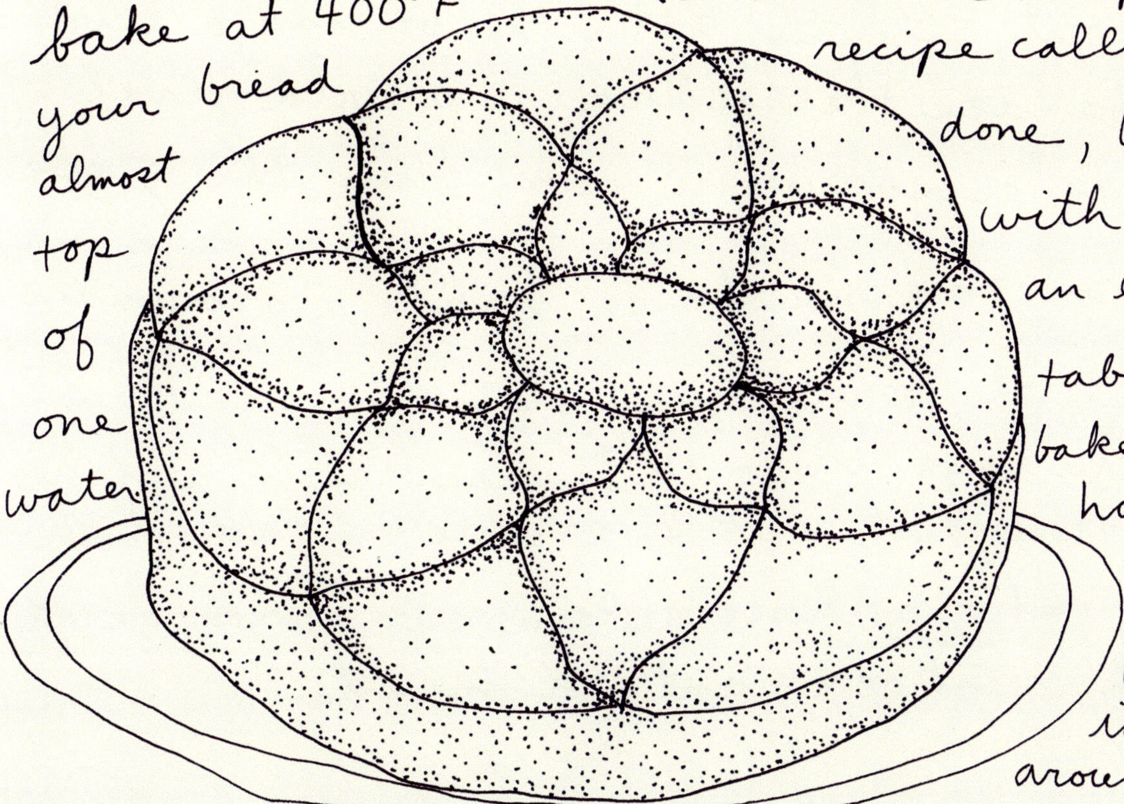

variations: roll a fruit or cheese filling inside each rope (& the ball) before braiding.

month of scorpio ♏ october 23 to november 21

birth ceremony. AUM with the mother during contractions. softly play tinkling drone music after the child is born; let him enter the world to the sound of angels.

death ceremony. dig a deep hole as long as the body of the dead person. place him in it naked and bury up to two feet below the level of the ground, and plant the favorite tree of the person over the grave. all this should be carried out silently. after the burial, all witnesses may abandon themselves to a wild revel of song & dance somewhere away from the grave.

moving in to a new house. make a procession going into the rooms, opening windows and burning cedar shavings or dried bay leaves in a frying pan or stone dish. AUM in a circle in each room.

moving out of a house. clean and arrange it with the next occupant in mind.

full moon in taurus.. meditate upon the center of the earth, on volcanoes & the creation of mountains, minerals & continents.

Onward Onward Ever Flow
D MODAL
A d minor
d minor7
d minor9
on-ward on-ward flows the wa-ter qui-et sounds and life be-low lap-ping
soft a-gainst the earth side ne-ver stop-ping e-ver flo-wing on-ward
on-ward e-ver flow
dmi. *
D major
find me in the mountain if I
sun-shine hum-ming mea-dows on the mountains e-ver
D major9
D6
go I ne-ver know find me smi-ling on the val-ley on-ward on-ward e-ver flow on-ward
down the mel-ting snow ri-ver run-ning in the val-ley on-ward on-ward e-ver flow (repeat A)
rocks are
stiff a-gainst the water ne-ver let their soft-ness show soon are
bro-ken down to san-dunes on-ward on-ward e-ver
flow (repeat A to *)
ABL '67

the month of sagittarius ♐ november 22 to december 21

harvest festival in the middle of a table assemble a bounteous tray of the most beautiful fruits and vegetables you have grown or gathered. surround the tray with paper harvest goddesses (see pages 110–111) and invite neighbors to bring their best culinary masterpieces to share. AUM together holding hands around the table and anyone can make an extemporaneous prayer of thanks for the bounty of the earth. after the meal more candles are lit, music & dancing begin & the table is cleared by everyone.

full moon in gemini meditate upon weather systems, air currents, wind & seasonal changes guitar: regular tuning

Thanksgiving Hymn

ABL oct.'69

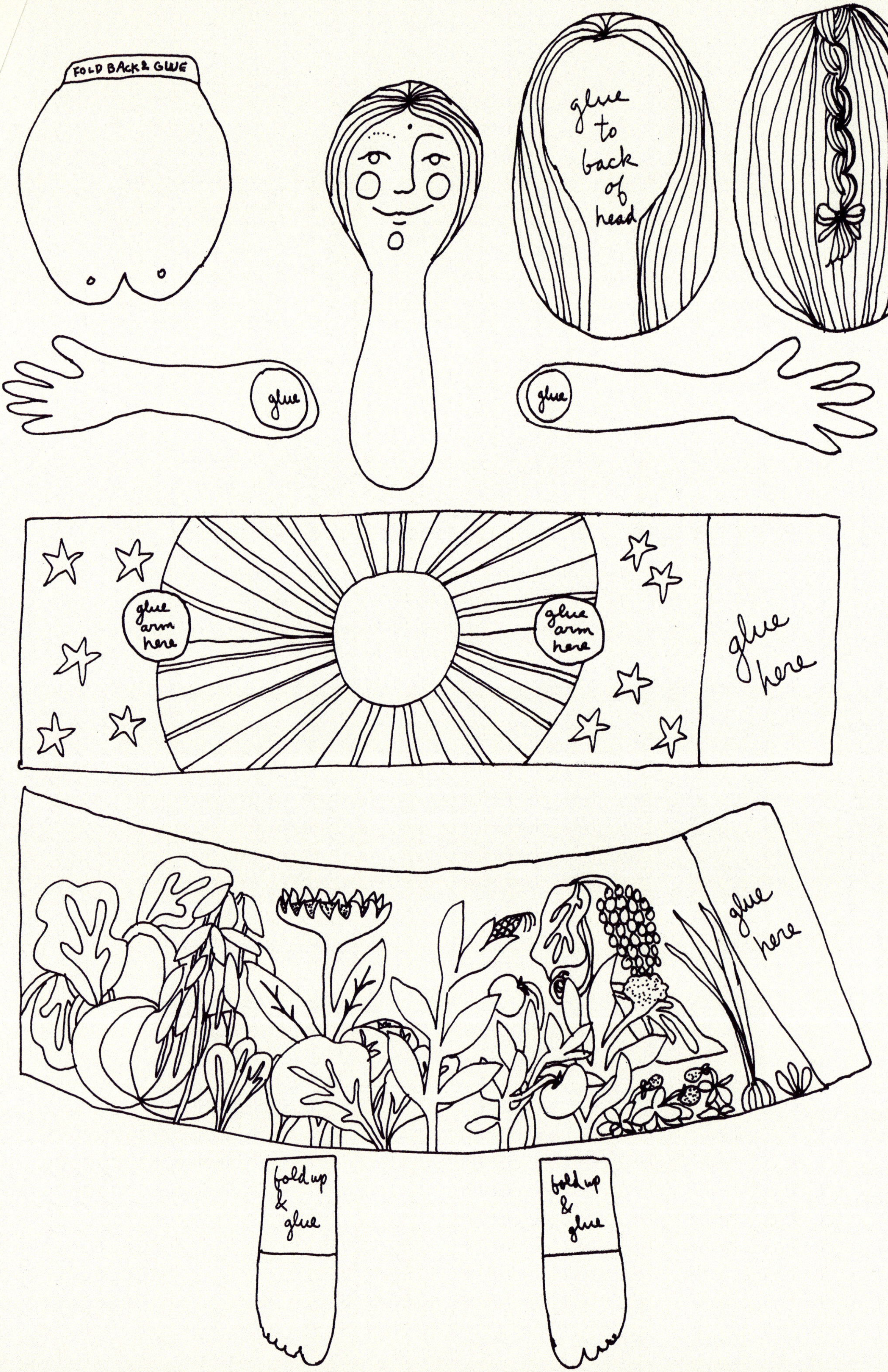
FOLD BACK & GLUE
glue to back of head
glue
glue
glue arm here
glue arm here
glue here
glue here
fold up & glue
fold up & glue

harvest goddesses to grace autumnal feasts are easy to make. trace these shapes and cut them out of strong white paper or colored construction paper. color them in or collage them with bits of magazine pictures. then glue them to an empty toilet paper roll as follows: first glue the blouse around the roll, lining up the top edge of the blouse with the edge of the roll. then glue the skirt around the roll, lining up the hem of the skirt with the other edge of the roll. (glue on the inside of the waist and on the glue tab only). then glue the arms to the arm sockets and the glue tabs of the feet to the inside of the roll under the front of the skirt. glue the face to the front hair and glue this united piece to the inside of the roll so that it faces the front of the dress. then glue the back of the hair to the front hair, matching the edges. then glue the tab on the bosom to the neck on the inside of the roll and gently tuck the ends of the breasts into the blouse & secure them with glue. You can put name cards or paper fruits or musical instruments in the hands. add wings for angels or a bouquet of straw flowers.

the month of capricorn ♑ december 22 to january 19

Winter Solstice Chant

winter solstice at midnight gather at a hilltop to make a bonfire if it is still, if windy, find a sheltered hillside or use an indoor fireplace if the weather is inclement. make a warm drink for all present (mulled wine, spiced tea, hot cider, hot spiced milk). much hugging and laughing and exchange of found or handmade things. at midnight sing the winter solstice chant and old favorite songs & carols. burn candles on metal plates in windows to guide travelers to shelter for this is the darkest night of the year.

full moon in cancer meditate upon the ocean & the flow of waters over the earth: lakes, rivers, rain, inland seas, & the blood of land creatures.

Receiver - Believer
regular tuning
C C7 F C C emi. ami. E7 F D9 G7 F C D9 G7 C emi. a mi. E7 F D9 G7 F dmi. a mi. D7 G7 C emi. ami. E7 F C G7 C
if you're loo-kin' thru the night for a place to take your ease a
if you're sing-in' on the paths of the gar-dens of the sun come
can-dle still shines from my win-dow if you're loo-kin' thru the world for a
sing a verse with me thru my win-dow if you're loo-kin' to the past for
gar--den of peace bet-ter stop your tra-vlin' on and look with-in no one can
glo-ry gone un-sung you've still to see the glo-ry be-gin tho' tho'
tell you a-ny-thing you don't feel in your heart you can get it straight to-day you're no
lon-ger a-part from the all ri-ver flo-win' the all gi-ver know-in' the
all re-cei-ver be-lie-ver re-cei--ver be-lie-ver
song of my soul sing it now sing it whole sing it all in a cir-cle sing with-
out and with-in all are one be-ing sun-beam be-lie-vers
ABL '71

the month of aquarius ♒ january 20 to february 18

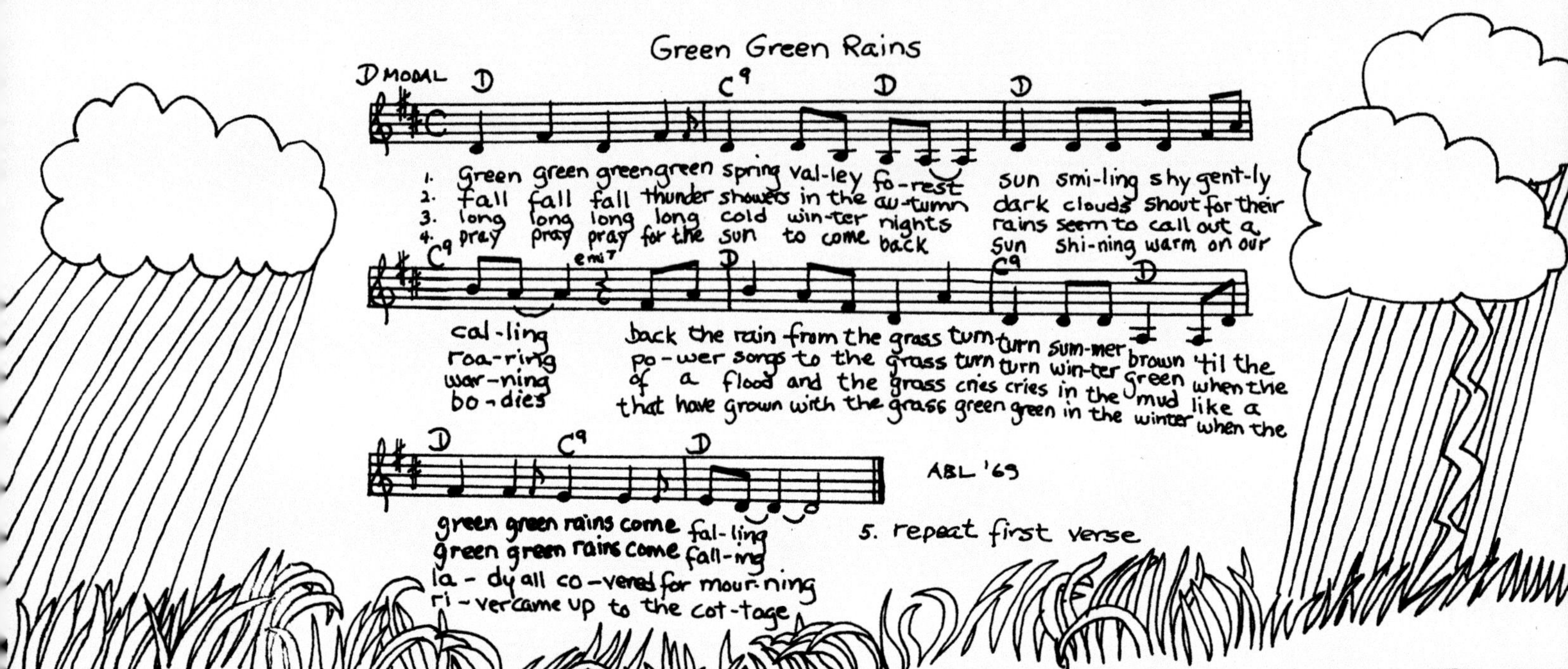

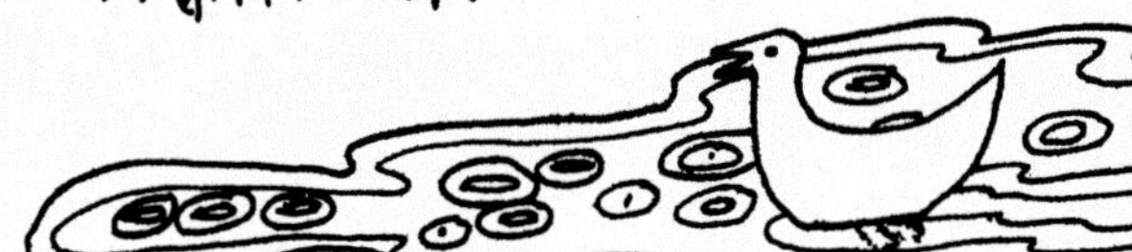

keeping warm if you are in a heated room, place a pan of water over the heater to keep your nose from drying out. winter health is improved by taking sauna baths often.

eclipse of sun or moon cut a small hole in a big piece of cardboard and let the eclipsing sun project through it onto a wall (camera oscura), & watch it happen on the wall. lunar eclipse can occasion an all-night meditation. eclipse chant:

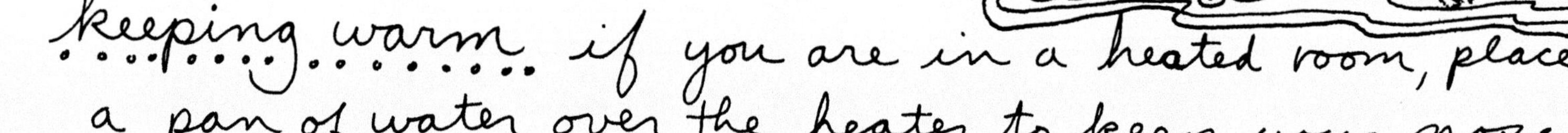

full moon in leo meditate on the sun & the light of the sun, in you, in all terrestrial life.

sauna bath

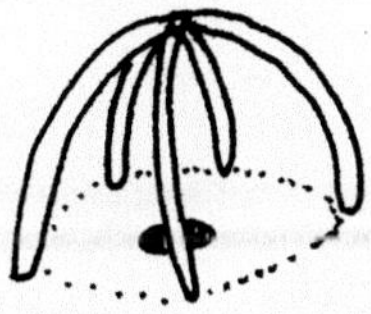

construct a dome 5½ feet high in the center and about 10 feet in diameter, either of curved branches buried one foot in the ground & lashed together at the top or a small geodesic dome. in the center dig a pit. near the walls spread hay & over that some cloth. cover the sides of the dome with construction plastic and insulating blankets and bury all the edges in the ground so that there are no leaks.

build a fire with large rocks stacked in with the burning logs. (volcanic rocks are best, igneous are better than sedimentary). after 3 or 4 hours of steady burning carefully dismantle the fire with a pitchfork and shake the hot rocks to get the cinders off, carry them into the dome & place them in the pit. fill 10 one-gallon containers with water & place in the dome for people to drink or wash themselves with or wet the rocks.

then everyone enters naked & finds a place to sit or stand (very careful of the hot rocks in the center) and the door is closed so no hot air leaks out. after the dry heat suffuses the dome, sprinkle a little water on the rocks. the steam is very hot! when the dome is cloudy with steam, place a small branch of fresh bay laurel, yerbabuena or mint or eucalyptus on the rocks. People may sing or chant or massage or wash each other or meditate on the steam.

one person may stand bent limp from the waist while two people stand on either side of him and drum up and down his spine with the little-finger sides of their hands until they are tired. then the person in the middle straightens up one vertebra at a time without speaking, feeling very tingling all over. (the other two should be prepared to steady him if he's dizzy). when each person has had enough heat, he leaves the dome, shutting the door tightly & takes a cold shower or a swim in a cold stream. they may re-enter the dome and repeat the process until they feel completely purified or the rocks are cold. after drying off, rub wheat germ oil on the skin.

caution: don't use river bed rocks as they can explode when heated.

the month of pisces ♓ february 19 to march 20

cloud-dispersing ceremony. (for two people)

tones

me	you
wind wind 5 1	tree tree 5 1
wind wind 5 1	tree tree 5 1
blow a-way 5 5 1	uncover me 1 5 5 1
all the clouds 5 5 1	in the sky 5 5 1
little (my name) 5 5 1	little (your name) 5 5 1

together:

want to see!
5 5 1

full moon in virgo. meditate upon the movement of the earth, on sand dust & deserts, the carving of canyons, the rock & roll of the planet.

elemental dances of abandonment.

earth dance run up a grassy hill or a sand dune and roll all the way down.

water dance float upon ocean surf. let it carry you to the shore.

air dance run around a windy meadow with an umbrella. let it lead you.

fire dance run naked in the sunshine.

rainy days... run naked in the rain.

we're all going to the same place; the difference is the time it takes.
don't mind Flow if her pace is slow;
we're all doing the best we can.

meditation & hatha yoga

mind is buddha, while the cessation of conceptual thought is the way. once you stop arousing concepts and thinking in terms of existence & nonexistence, long & short, other & self, active & passive, & suchlike, you will find that your mind is intrinsically the buddha, that the buddha is intrinsically mind, and that the mind resembles a void. only come to know the nature of your own mind in which there is no self and other, & you will be in fact a buddha.

huang-po
(9th century zen master)
translated by
john blofeld

the mind

to use the full capacity of one's brain is to experience simultaneously everything happening in the here & now, from the cellular activity of the body to the farthest reaches of the universe. sensory perception is mental, but it is not thinking, which is talking to oneself without making any noise. thought filters out sensory awareness. meditation is a way to rise above the thought-maker into the here & now, the realm of bliss.

breathing

to begin a meditation relax your breathing. feel your asshole. when you inhale, your asshole should expand (relax) and when you exhale it should contract (draw up).
breathe through your nose rather than your mouth, and breathe from your diaphragm (belly) rather than your chest.
observe the rhythms & sounds of your sleep breathing as you wake up, and copy them in meditation & daily living. the sound of your breathing should come from your chest & throat rather than your mouth or nose. try out some long gentle sighs.

breath counting technique

each inhale or exhale counts one number. you can do it a variety of ways: our favorite is

in	out	in	out	in	out	in	out	in	out	in	out
1	2	3	4	5	6	7	8	9	10	9	8
7	6	5	4	3	2	1	0	1	2	3	4

and so forth. this concentration exercise may be enhanced by the visualization

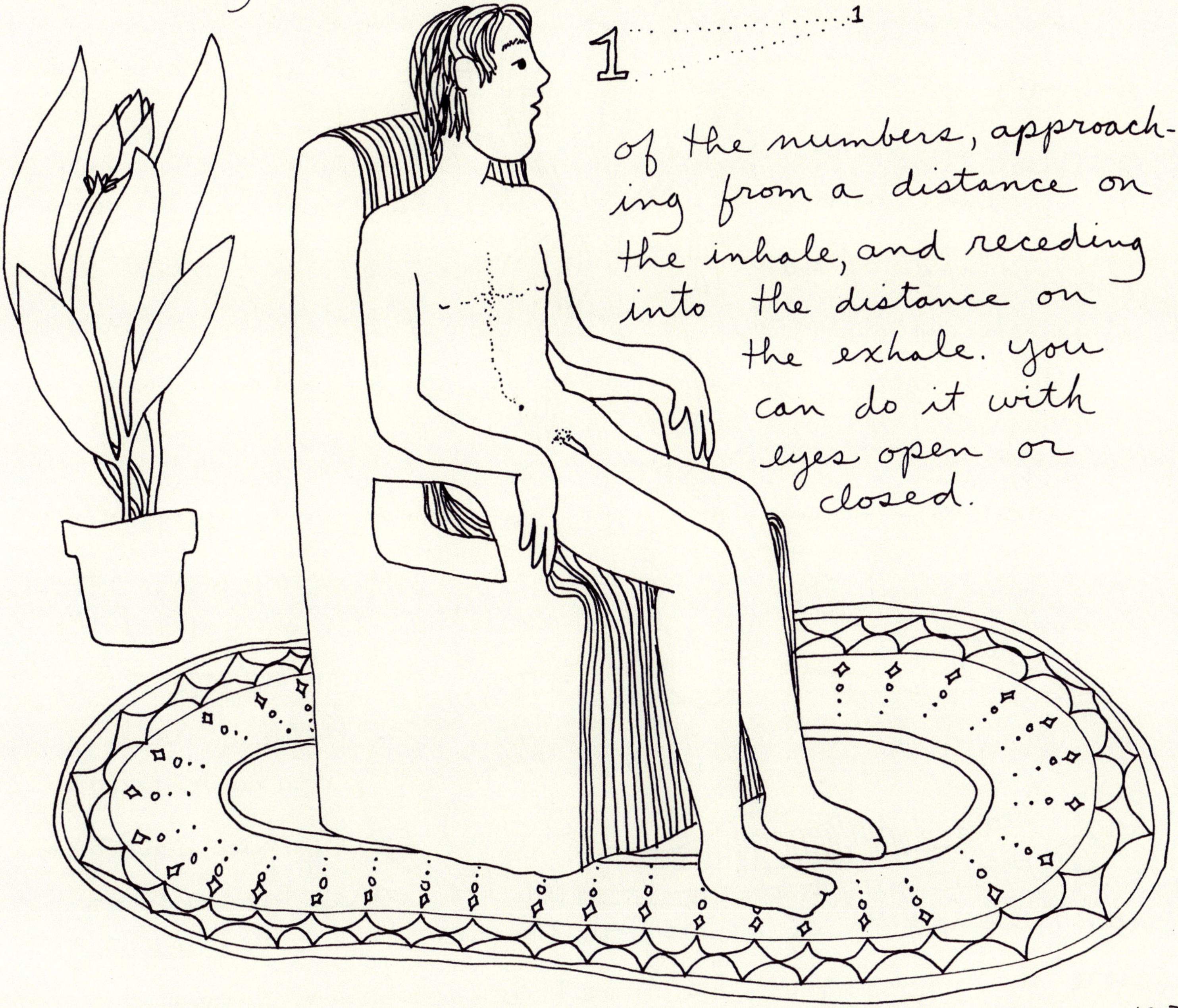

of the numbers, approaching from a distance on the inhale, and receding into the distance on the exhale. you can do it with eyes open or closed.

energy currents

as you relax your breathing you begin to feel your whole body pulsating. then you become aware of currents of energy within the body. Usually it starts as a tingling sensation in the hands or feet, spreading in waves through the whole body. The flow of this energy (called prana in sanskrit) can be directed at will throughout the body. You can "breathe" it in through your nose and out through your toes or in through your toes and out through your nose. You can breathe it in through the palms of your hands and out through your heart or the center of your forehead. You can breathe in a figure 8, inhaling down your spine & curving forward into your belly, and exhaling around your crotch, up the lower spine & out through the chest. the early hours of the morning are the most tranquil & full of invigorating prana. the hours before midnight give the most refreshing sleep.

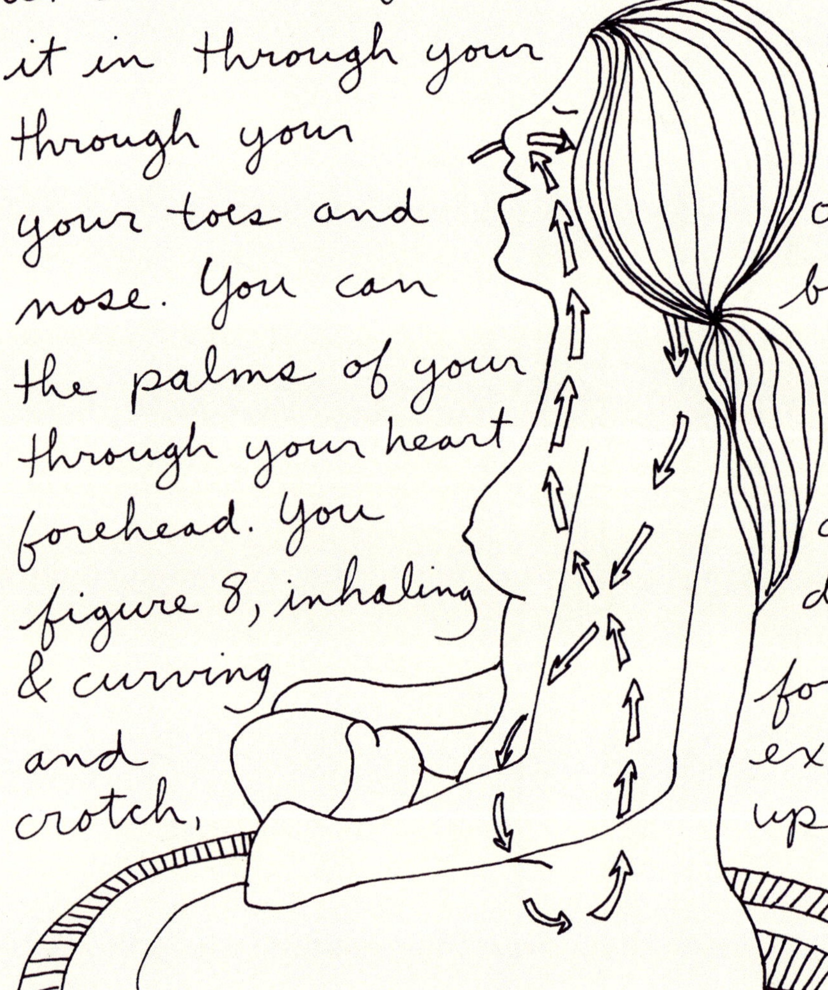

the eyes may fix relaxedly on a candle flame, a mandala (design flowing out from a center), the moon, a dried sun-flower center, a drop of dew, a fragrant blossom, a star, the eyes of another who is gazing in your eyes, or the void. (or see the sunlight meditations in this book).

to contemplate the void: fix your eyes on a point in space 4 feet in front of your face.

positions for meditating:

sit with your spine straight and shoulders relaxed. rock your body back & forth until you find a comfortable place to stop. you can sit cross legged or in a chair or in lotus or ½ lotus. you can wrap cloth around your knees and the small of your back. tibetan monks make wooden boxes to brace the back & the knees. zen monks sit on firm round pillows. you can also sit on your heels.

you can meditate while walking: neither stiff nor slouching, balancing firmly on both feet, each step an exhale or an inhale as slowly as you can.

you can meditate lying down, but most people who try it fall asleep. mudras or hand positions will help you stay awake:

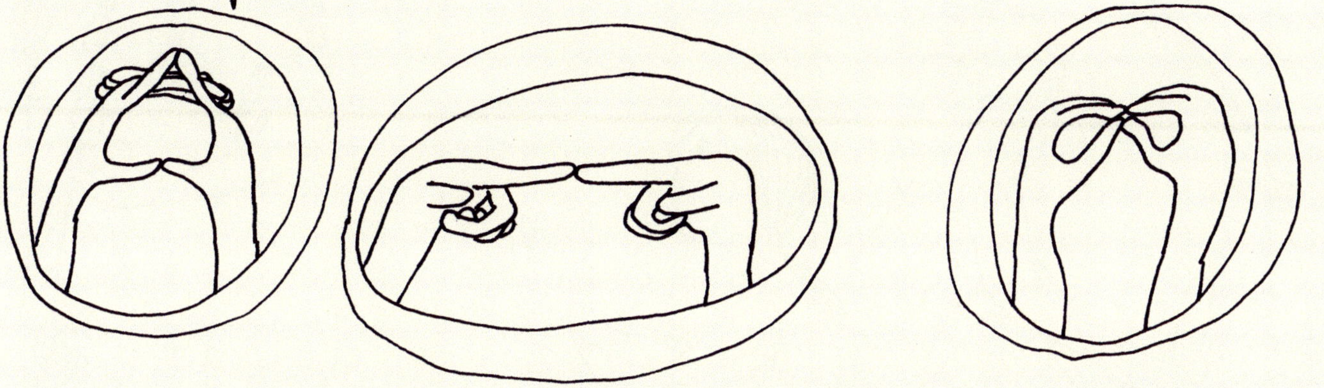

this tibetan hand mudra is traditionally done standing on one foot with the sole of one foot against the inside of the thigh of the other leg, but you can also do it on your back:

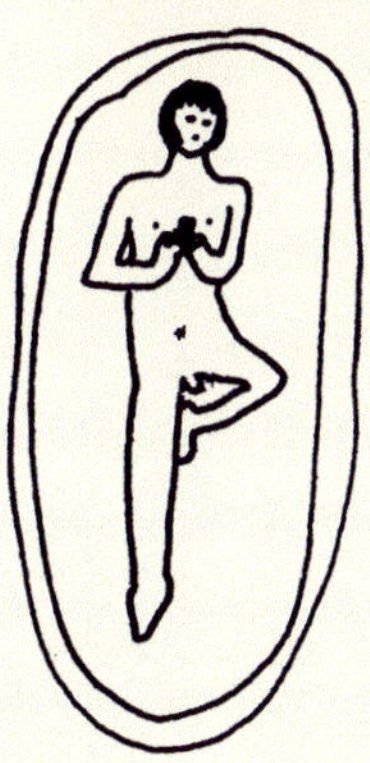

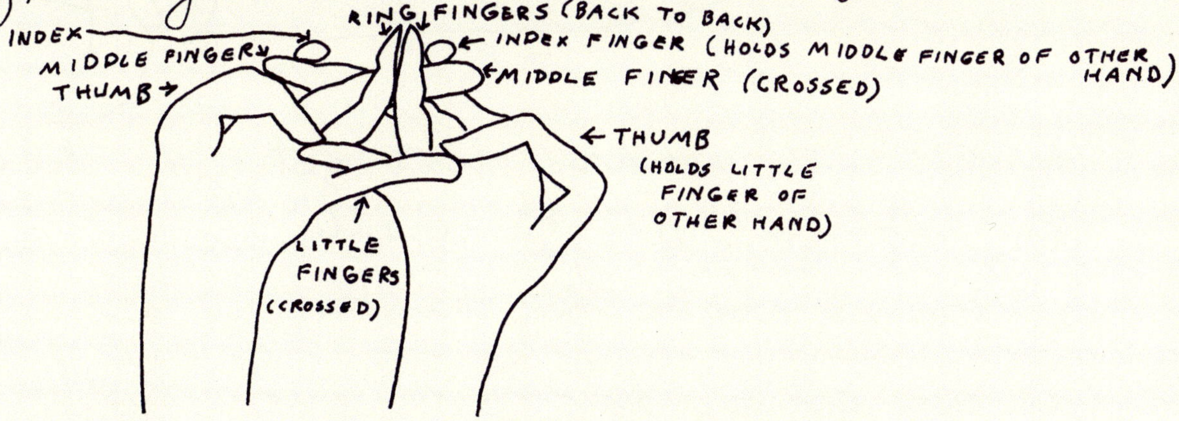

you can make up your own hand mudras. this one is our favorite for lying on our backs in a grove watching the light come through the leaves:

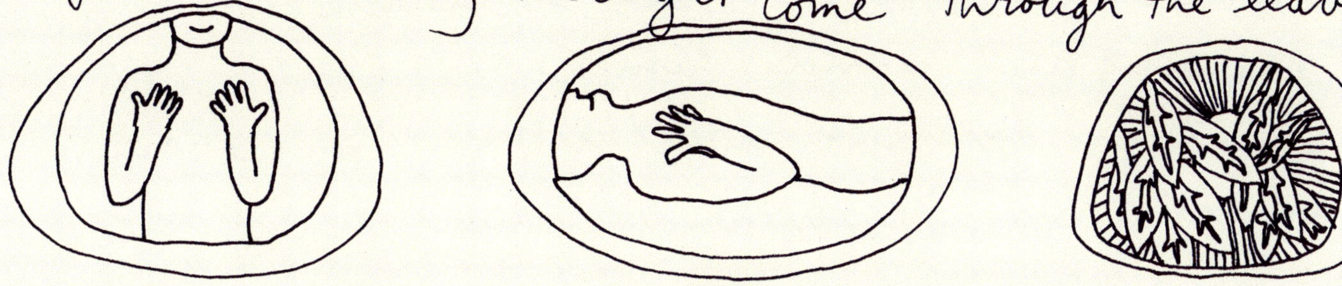

mantra

a word or words that get you high can be repeated rhythmically: as a method of rising above the thinker, as a meditation, as an act of loving devotion to the one, as a greeting, or as an accompaniment to dancing or other mindless activity. drums, finger cymbals, drone instruments, hand clapping and wind chimes are good with mantras. You can breathe them, sing them, whisper them, speak them or chant on one tone.

speaking mantras

BOM SHANKAR BOLENAT "thank you, god, for reminding me that I am the Universe." (a good greeting)

MAY ALL BEINGS BE PEACEFUL & HAPPY FOREVER (say it 6 times - once to the earth, once to the sky and once in each of the four directions.)

breathing mantras

Shanti (peace) inhale "shan" exhale "ti"

So ham (I am He) inhale "so" exhale "ham"

singing mantras with melodies

this means, in sanskrit: "all evil vanishes from him who keeps the sun in his heart".

........this one goes well with fiddle, dulcimer & jaw harps

singing mantra on one note

the gayatri (at the beginning of the hatha yoga pages) is one of our favorites. begin with an AUM and then sing the rest on the same note as the AUM. when repeating it, sing an AUM between each verse.

AUM

is a name for the all in sanskrit and is the source of the latin word omni, meaning all, also the word amen. it is a chant of open & closed mouth humming on a single tone. begin singing

ah (as in f<u>a</u>ther), then O (as in b<u>oa</u>t), then hum mmmm. each AUM takes one long exhale. You can hear the harmonics on the tone you're singing as you move your lips. a room filled with hundreds of people AUMing sounds like a tremendous bell. besides being an effective meditation mantra, AUM is a fine way to say grace before a meal or conclude a meeting of close friends. AUM together holding hands.

an altar is a meditation environme

here are the elements we like to include:

mandala

sunstrobing provides many designs for these circular & symmetrical shapes flowing from a center. draw them often and you will find one that really draws your gaze, which can be used as design for an embroidery, painting or prayer rug. Inward curling spirals turning on a wheel are good. (see sunstrobes)

dish of water

light reflected in water is relaxing to look at.

incense

keeps time (five minutes or more per stick). keeps insects away. provides a point to gaze on, odor makes a tranquil mood. place a dish underneath to catch the ashes.

candles

illuminate altar when surrounded by darkness, provide points to gaze on. candlelight, like sunlight, is easier on one's nerves than electric light.

bowl gong

or a resonant bell or flat gong is a bringer to the here & now. the decaying sound can be centered on when beginning meditation. also good for punctuating singing and chanting or the beginning and end of a session of meditation or hatha yoga

natural things

for their good vibes. sea shells, potted plants, dish gardens, smooth stones.

a comfortable place to sit

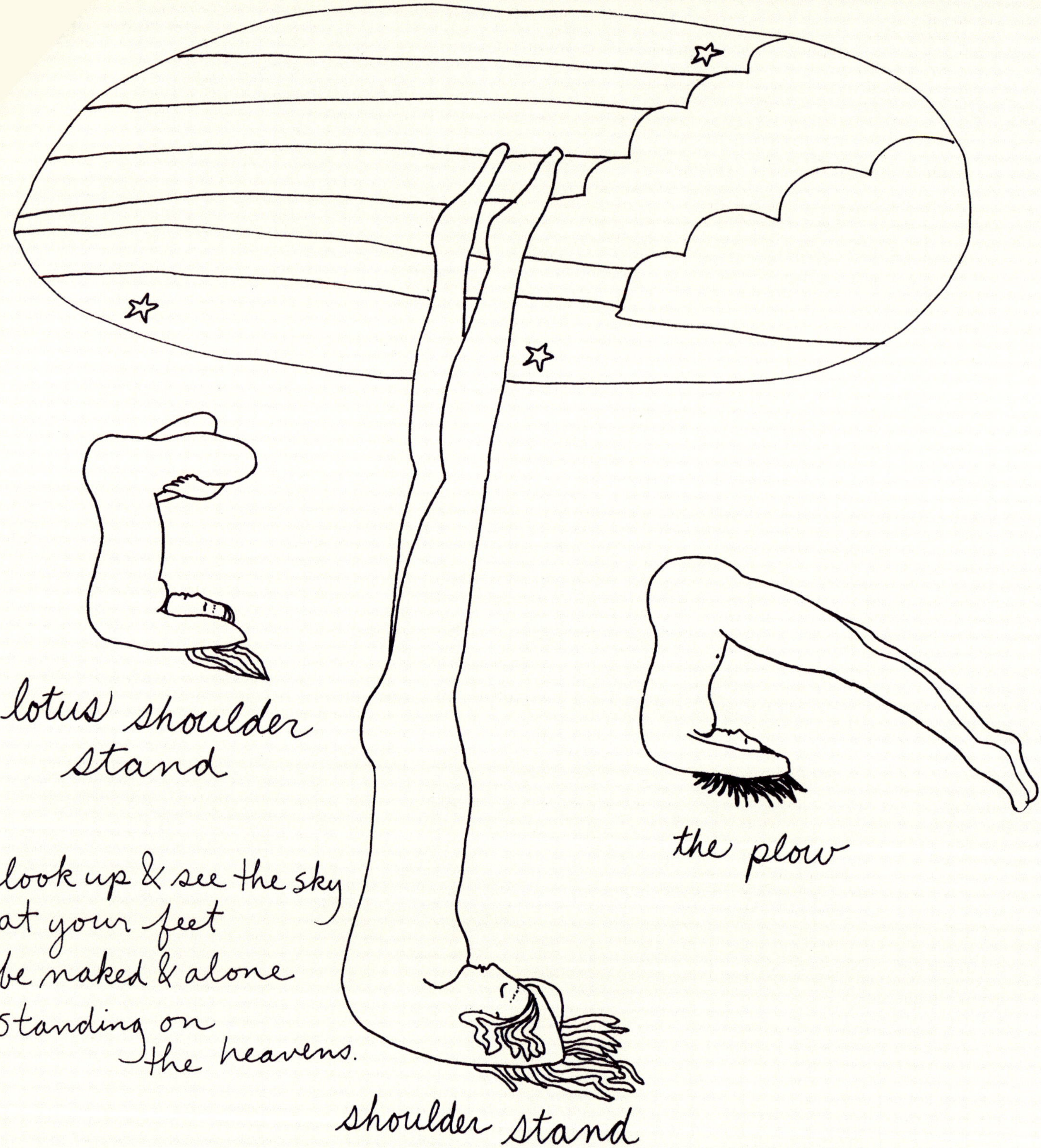

we are not teachers of hatha yoga, but we submit our routine as an example of what can be developed for personal use. if you try them, please do them slowly & gently, resting between each posture. the aim is not how many you can do, but how many relaxed breaths you can hold each one without straining. If you feel sore afterward you've been pushing too hard. Take it easy!

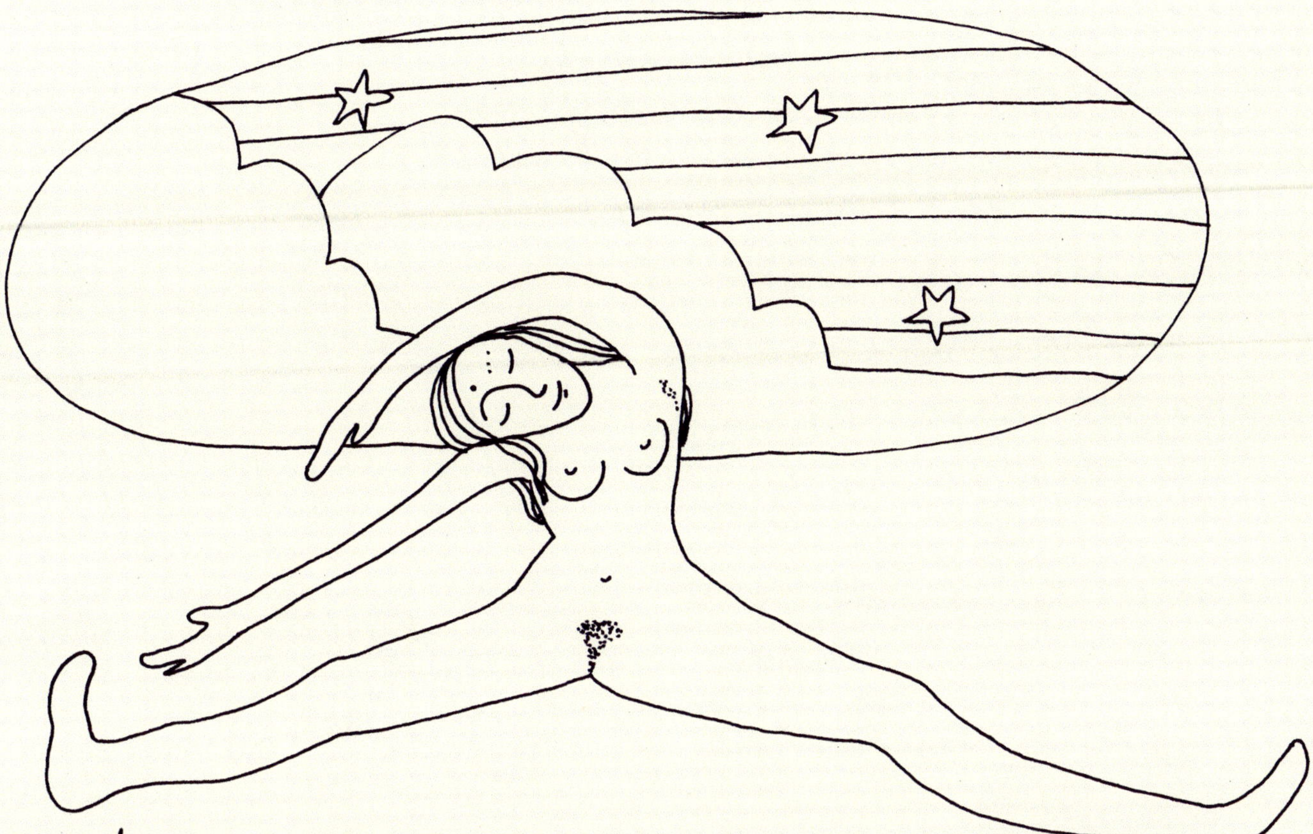

the best time to do the asanas (postures) is in the morning before breakfast, after having taken a shit and if possible a bath or a swim. the best place is on a flat, clean, firmly padded surface, such as a folded blanket – if possible do them nude outdoors in the cool sunshine or warm shade. a morning like this makes us feel good all day. Hatha yoga strengthens the back, quiets the mind, stimulates blood circulation, calms nerves, takes off excess flabbiness & limbers up the whole body. It enables you to sit or stand without leaning on anything. If you have a hatha yoga teacher watch & correct your asanas it will save you much time & struggle. Rest between poses as many breaths as you spent doing the last one.

our hatha yoga sequence

the gayatri mantra is one of the oldest hymns known to man. it is chanted to the rising sun. below is ramón's translation from the Sanskrit. sing it on one tone several times, slowly:

AUM Bhûr Bhuvah Svãr
Tat Savitur Varenyam
Bhargo Devasya Dhimahi
Dyo yo Nah Prachodayat

AUM oh earth, oh air, oh golden light
oh that sun, most adored
we drink the splendor of that God
may he fill our day with light.

eye exercises

A. alternate looking up as far as you can and down as far as you can for several relaxed breaths.

B. then alternate from side to side as far as you can for several relaxed breaths.

C. alternate upper right to lower left and upper left to lower right

D. roll eyes clockwise and counter clockwise.

surya namaskar (greeting to the sun).

If you have time to do only one exercise, do this one. The 12 poses are the months of the year.

A. face sun with palms together in greeting. breathe out.

B. breathe in as you lean back with palms over head

C. touch the floor in front of feet (breathe out)

D. breathe in as you bring right leg forward, left leg back.

E. hold breath as you bring right leg back.

F. breathe out. touch floor with forearms, hands, forehead, chest, knees and toes

G. breathe in as you rise up with legs, hands & stomach on the floor

H. repeat E breathing out.

I. repeat D. breathing in with left leg forward

J. repeat C. breathe out

K. repeat B. breathe in

L. repeat A breathe out

STANDING ASANAS

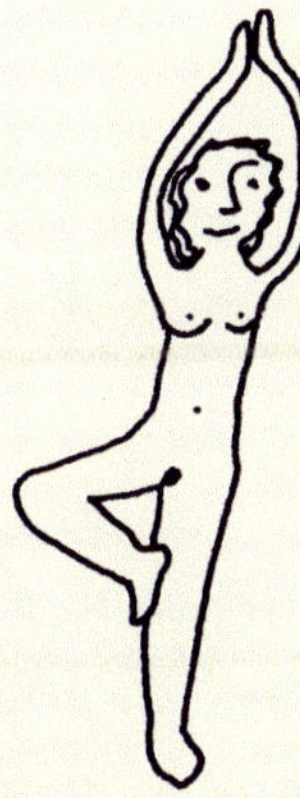

the mountain:
stand with weight equal on heels & balls of both feet. contract belly and relax shoulders with spine straight. Observe slow breathing.

the tree:
start in the mountain pose. lift one leg so that sole of foot is on inner thigh and knee goes out to the side. Place palms together over head with arms straight. Observe breath.

additional: these three poses before returning to the tree and then the mountain. then do the whole pose again on the other foot.

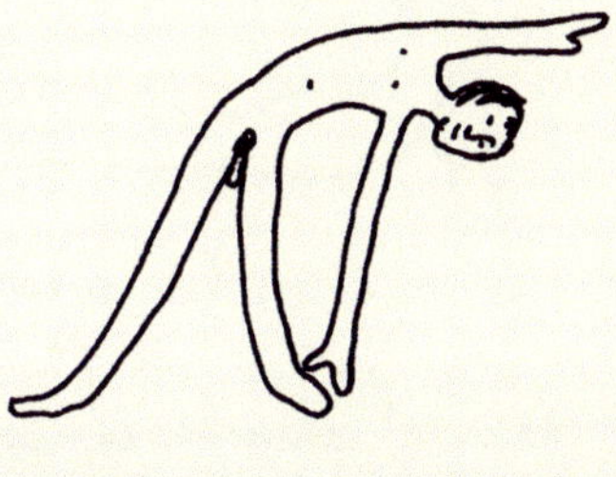

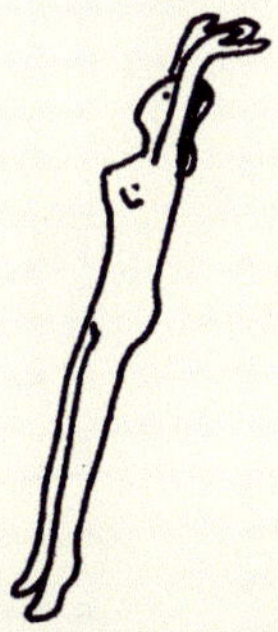

stretches: with knees straight bend over to one side with arm outstretched and touch foot with other hand. observe slow breathing. do both sides.

ceiling stretch: rise on tiptoes and stretch whole body up with palms parallel to ceiling. hold for as many relaxed breaths as you can.

the cow: touch palms to the floor (or ankles) and hold as many relaxed breaths as you can.

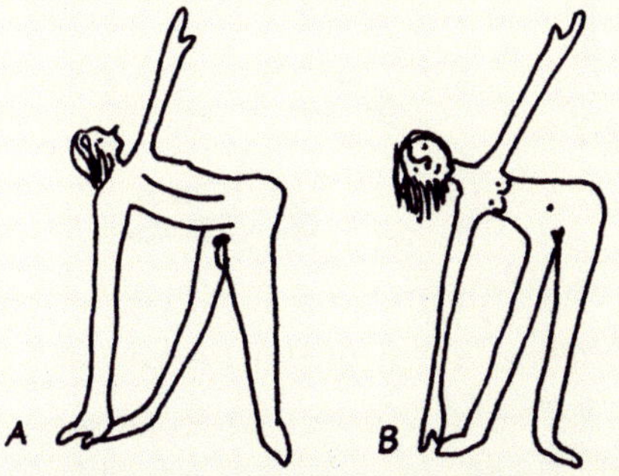

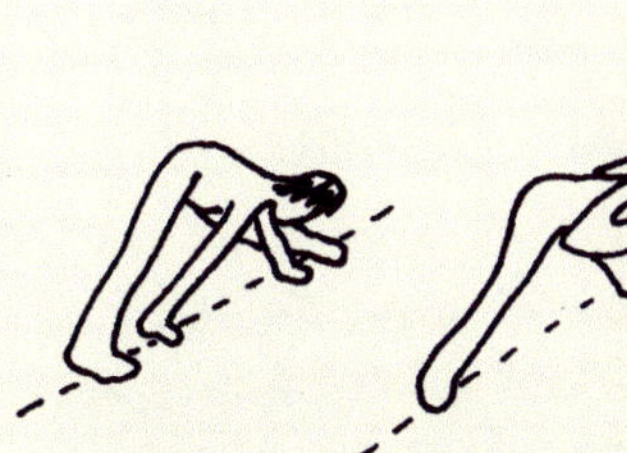

triangle toe touches:
spread legs with knees straight. A place right hand on left foot, extend left arm above head & look at it. B place left hand on left foot and look at right hand above head. Then both poses on right foot.

extended foot pose.
spread legs with knees straight and feet parallel. place palms in line with feet between them. If easy, place palms together with fingers toward head behind your back. Observe slow breathing. . .

rooster pose:
Stand straight with palms together behind back, fingers toward head, and place one foot a step in front of the other. A: stretch head to knees. B: stretch head towards buttocks. change feet. observe breath.

flying: stand straight with palms together, arms straight over your head. if you do this one slowly you can get several relaxed breaths in each pose. A: stretch upwards B. lunge forward onto right foot C. stretch back flat, shift weight onto right foot. D. balance on right foot with arms extended forward and left leg extended straight back. E. stretch arms out from sides. when done perfectly the head, palms, left foot, knee, and the whole back are the same distance from the floor. F. return to standing position. do it again on left foot.

LYING PRONE & ON BACK

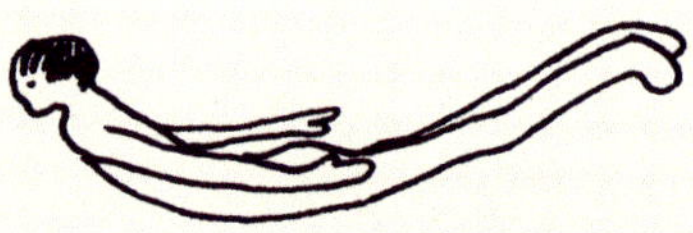

the locust: lie prone with arms at sides. Then lift legs, arms & head, without bending elbows or knees. hold several relaxed breaths and then lie prone again.

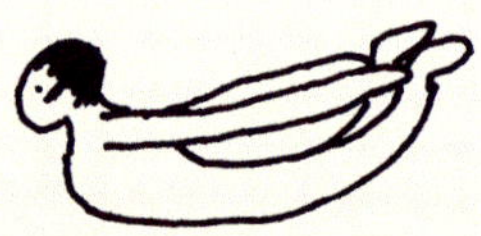

the bow: lie prone with arms at sides. Then lift head and graspankles with hands. (you can bend your knees). hold several relaxed breaths and then lie prone again.

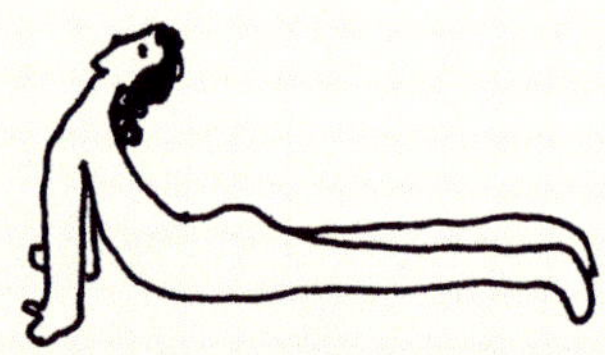

the cobra: lie prone with palms on floor beside shoulders. look up, bend head back & lift up chest using arms as little as possible. hold several relaxed breaths and then come down head last.

anantasana: named for the serpent upon whom the god vishnu sleeps & dreams the universe. lie on one side with lower elbow outstretched & palm on ear. stretch up the upper leg and clasp foot with upper hand. hold several breaths. then do the other side. try not to bend your knees.

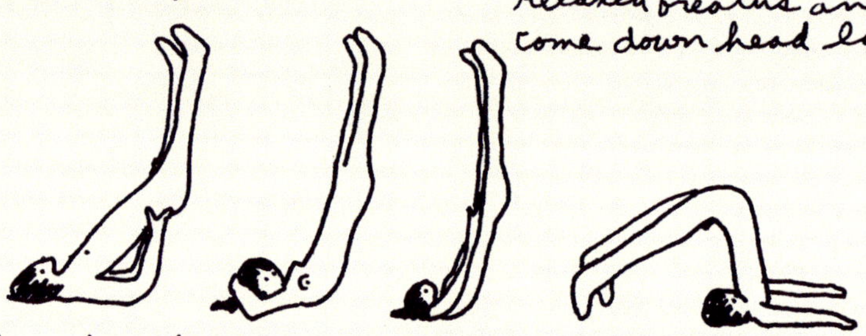

the shoulder stand and the plough: lie on your back with arms at your sides. bring your legs & back up so that all your weight is on your head and shoulders. your arms can be in either of 3 positions shown above in order of easiness. hold as many relaxed breaths as you can. Then bring feet over your head to the floor (legs straight) and stretch arms out straight on the floor behind your back. hold this pose several relaxed breaths. then bring your feet straight up again and come down slowly with legs straight, one vertebra at a time. you may feel a satisfying crack or two as your vertebrae are on the floor and you slowly lower your legs (don't bend knees).

KNEELING & SITTING

A 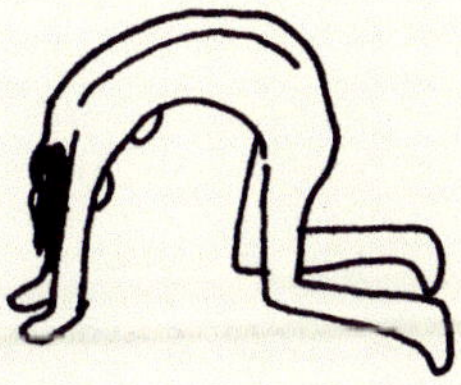 B  C  D 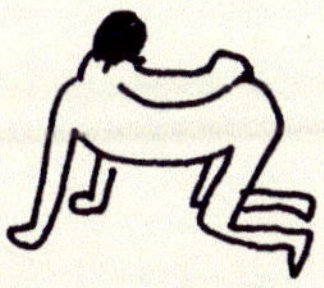

get loose: stand on hands and knees. A: lift up back, lower head and bottom, breathe out. B: lift up head & bottom, lower back, breathe in. alternate these two for several relaxed breaths. C bring hips & head towards each other on the left side with back at normal level. breathe in. D bring hips and head towards each other on right side, breathe out. alternate these two for several relaxed breaths. Then do a whole circle: A-C-B-D for several relaxed breaths. Then reverse direction.

the lion: cross one bent knee over the other and place both hands on knees arms straight. Open your eyes as wide as you can. Open your mouth as wide as you can. Stick out your tongue. hold several relaxed breaths. good for sore throats!

the camel: sit on your calves. Then rise up straight from your knees, place hands upon heels and lean back head. hold several relaxed breaths and return to kneeling or go into the fish position (next).

the fish: from a kneeling or camel pose bring palms together on chest and bring head back to the floor. hold several relaxed breaths. Use back to help yourself up to kneeling position.

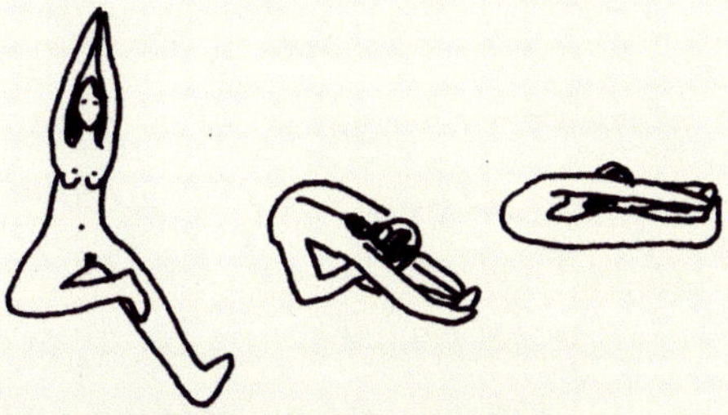

head to knee stretch: sit with one leg straight out and the other bent so that the foot rests on the thigh (or sole against inside of thigh) palms together stretch spine up, then stretch head to knee and clasp foot (or ankle). hold several breaths. Then do other leg. then stretch to both knees.

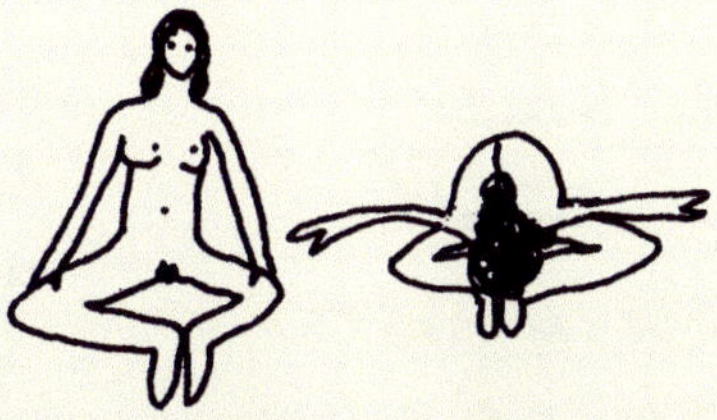

the cobbler: sit with knees bent out and soles of feet together. bring feet as near to body and knees as near to floor as possible. Keep spine straight. hold several breaths. Stretch head to feet. hold several breaths. return to sitting.

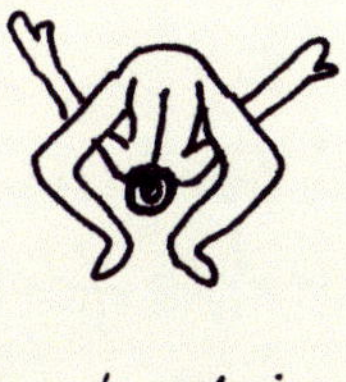

the tortoise: sit with knees slightly bent bring head, shoulders to floor between legs and thrust arms back under thighs. hold several relaxed breaths and sit up.

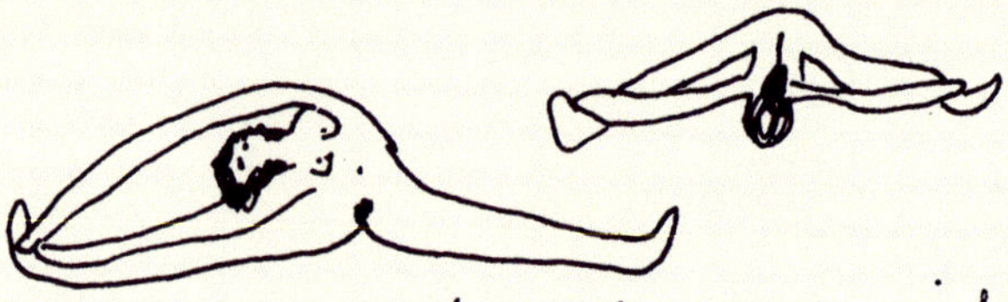

open stretch. Sit up straight with legs straight and feet as far apart as they'll go comfortably. stretch over to one ankle holding foot in both hands. Then up straight and do other side. Then hold one ankle(or foot) in each hand and bring head to floor. hold each position several relaxed breaths.

spinal twist: sit with right knee bended under left thigh and left foot outside of right thigh. hold left foot in right hand with arm straight & elbow on left side of right knee. bend left arm around back with hand on waistline as far around as you can. hold several breaths. do it on the other side.

the lotus and the half-lotus: sit with spine straight and belly in with legs crossed indian style. Place one foot on the thigh of the other leg. (half lotus). Place other foot on thigh of first leg. (lotus). hold whichever is comfortable for as many relaxed breaths as you can.

another lion: from lotus Come forward so you rest on your hands and knees. Open eyes and mouth as wide as you can. Stick your tongue out. hold several relaxed breaths.

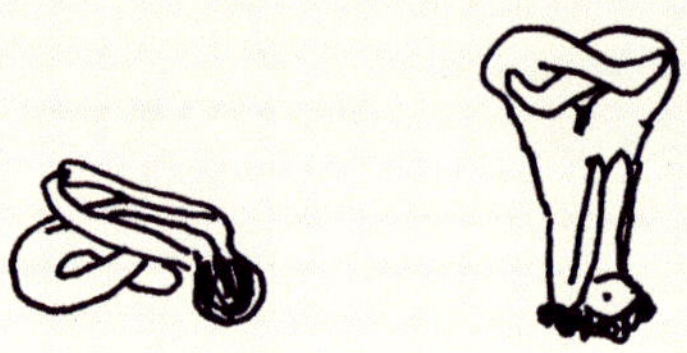

yoga mudra: the symbol of union. Sit in lotus. join hands behind back with arms straight. bring head to floor. hold several breaths. also you can do a shoulder stand in the lotus position.

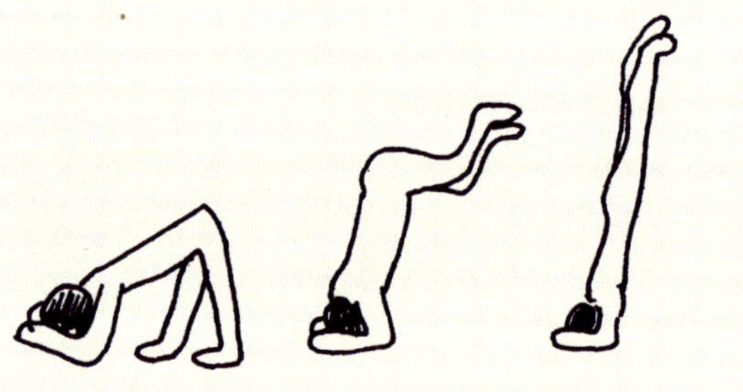

head stand: make a triangle of your forearms and the space between your elbows (all sides equal.) rest your head between your clasped hands. with your legs straight, "walk" up to your head until your back raises you up and your legs straighten. hold several relaxed breaths and come down the way you came up.

rest: at the end of asanas rest on your back with your palms up next to your thighs for many relaxed breaths.

environment

away from civilization one need only quiet one's mind to be in paradise. when you move there, though, be sensitive to local social codes, so that your public appearance makes you blend in harmoniously. no need to blow your neighbors' minds, even if cosmic experiences move you to change your own life.
140 and leave no tracks. leave your campsites as if no one had ever been there. respect people's privacy.

the ideal human climate is one where clothes, shoes, houses, heating, cooking are not necessary. our species lived that way before the first civilization. if you can raise what you like to eat, you're probably in your own ideal climate. All you need is sense enough to go outdoors when the sun is shining.

There is no doubt that a tree makes the best house. the most tranquil room combines natural and handmade useful and beautiful objects, green growing plants & emptiness in a light and airy space constructed of natural materials, such as wood, stone, thatch, bamboo and adobe. Ideally each bed should be in its own structure, out of eyesight & earshot of other structures. a round house, a tipi or a dome makes an indoor space more like the universe and less like a box. floor level sitting encourages a stronger back. No electricity puts you on a healthy solar rhythm. living areas are

defined by your day: sleeping & meditation (when it's inclement weather), preparation of food, bathing & laundry, playing music & arts & crafts, dining & sitting around

we like to hang cut-out paper stars from the ceiling because walking through them is like walking through the heavens, and because breezes turn them so they disappear (when they're sideways to you) & twinkle, and paper moons go through phases. colored cut-out rainbow lady smiles blessings 'round the room. use stiff paper. take needle & thread to make a hole at the top & tie thread through it. tie other end to tape, tack or through a canopy.

to start a woodfire: hold a magnifying glass over some cumpled paper, straw & thin kindling wood. the sun's rays, concentrated by the glass, will eventually cause the kindling to smolder. Fan it gently, and when it is flaming, gradually add bigger pieces of wood.

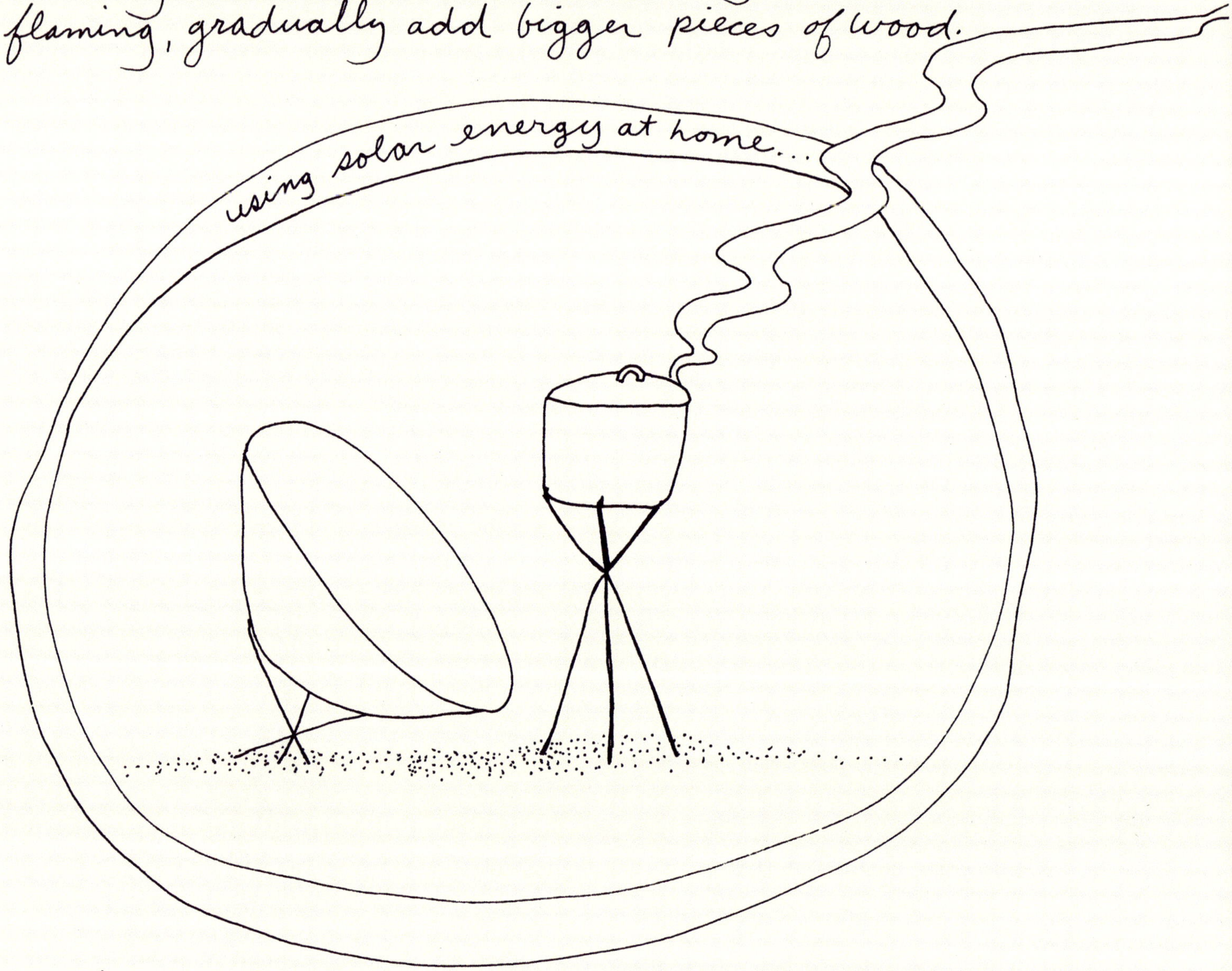

parabolic mirrors: (above) the sunlight is focussed into the center and to a point above the center, where a cooking pot, water heater or oven may be placed. a solar furnace for high temperature metal work utilizes many parabolic mirrors all focussed on the kiln.

easy water heater: coil 200 feet of 3/4 inch black plastic hose running from water source to bathing place, either across a sunny meadow or on the roof.

communities fall generally into two lifestyles: those with one kitchen for everybody and those in which each house has its own. communities come in so many varieties that people can pick the most harmonious with their individual lifestyles.

the purpose of the intentional commune is to provide as much leisure as possible for all its members. some leisure-preserving precautions for communal groups:

- don't freak out the locals.
- don't encourage visitors.
- don't exceed the number of residents your land can support.
- pay attention to the land & weather: feed the soil, pick up the garbage, keep cars & polluting chemicals off it.
- no leaders. instead, a tribe of volunteers.

communal living is the best for children because there are many kids and adults to relate to.
parents can take turns watching them, leaving others free to be other places.
communes focused on their own schools are the best.

possessions

as needs are simplified & communal trust blossoms, ownership gives way to usership; unused things are put in the community's free store to be recycled into another house-hold. In a magical way, everyone finds what he needs, & money becomes less & less necessary. when moving out of a house, each person leaves what he can for the next person & travels light. artists create beautiful things just to watch them circulate in the flow of usership.

relationships

living on land held in common brings deep relationships with others living there, often lasting for lifetimes. Sometimes it is hard to find time to be alone, because everyone who drops by is one of your closest friends. a sign on the door or a verbal request sometimes fills the function of a telephone. When several dwellings are on the same canyon, it acts as an amplifier & echo chamber for long distance communications.

canyon calls

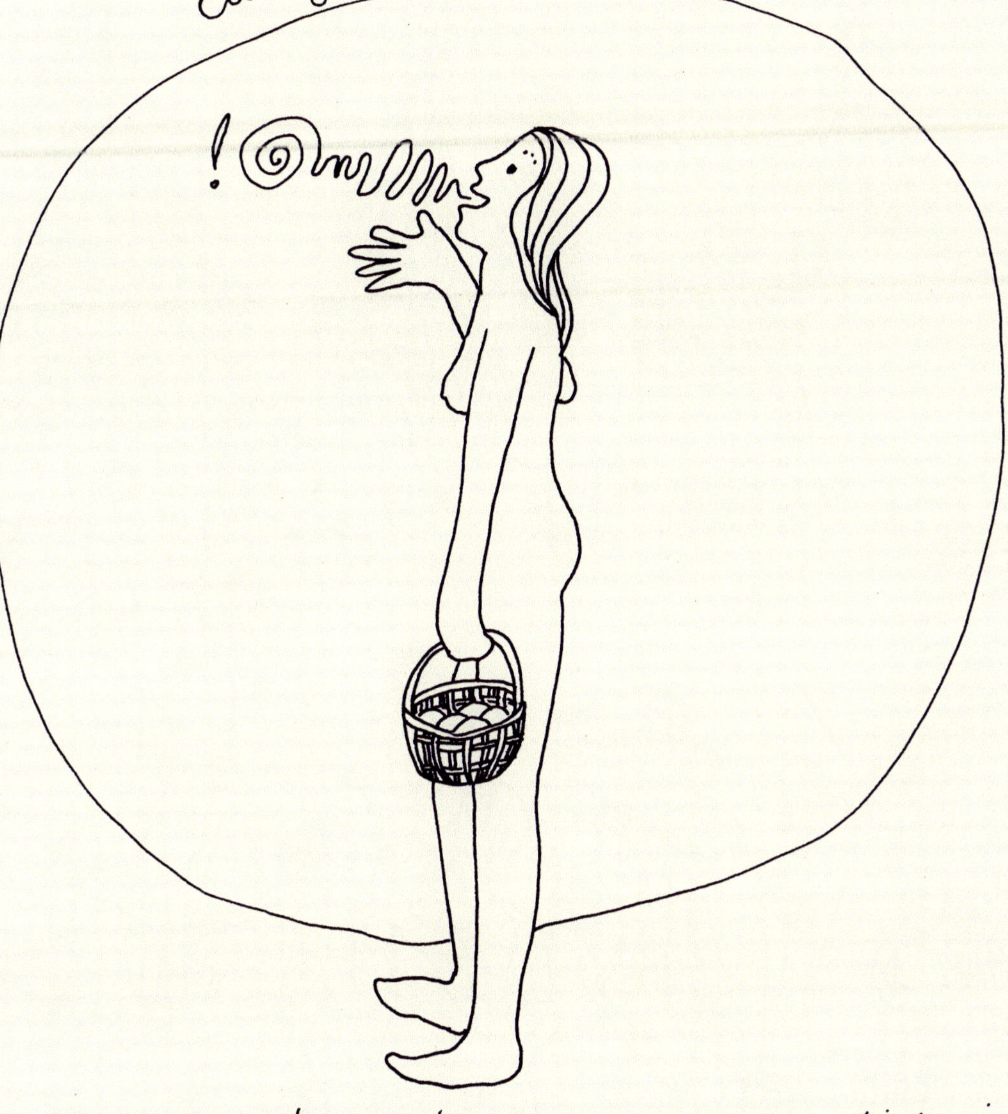

each person has his own identifying call: a yodel, a hoot, a cheer, a whistle, a laugh, a song, a howl, a moo, any animal's call or any kind of funny sound. distantly approaching an encampment, a visitor lets out his personal canyon call to announce who is coming and the replying call will tell who is there to welcome him. Silence tells that no one is there or that no visitors are wanted. a canyon call can also be a mantra (a word or words inducing god-consciousness).

sunday feasts are
pot luck dinners
with everyone both listening & performing happy music together & dancing. a sauna bath or swimming or watching a lunar eclipse or celebrating birthdays & weddings & visiting gurus go well with sunday feasts. each one is different, but all are wonderful because all are hosts and all are guests.

work day parties

are gatherings to do a project; preparing the soil for a garden, raising a roof, collecting & recycling garbage, building a dam are more fun with music & lemonade. morning & evening milking sometimes draws singers & bakers of hot biscuits.

meetings for discussing community business can be run with best results when only one person talks at a time. A stone or a shell can be passed around the meeting circle; whoever holds it may speak; if they have nothing to say they pass it on. In a community of leaderless volunteers, every voice carries equal weight.

sometimes in their each of our groups of people get so high & so close interrelating that they feel at one with each other and with the universe. some of our group religious experiences came out of these events:

a feast in which everyone fed everyone except themselves (with their hands)

sleeping in a circle with all the feet in the center, alternating men & women.

witnessing together a birth, marriage, wake, eclipse, sunset, sunrise, meteor.

working together at thankless tasks.

chanting AUM while hugging in a circle.

singing & dancing & playing music together with abandon.

taking a sacrament.

neighborhood silence fast.

living on open land.

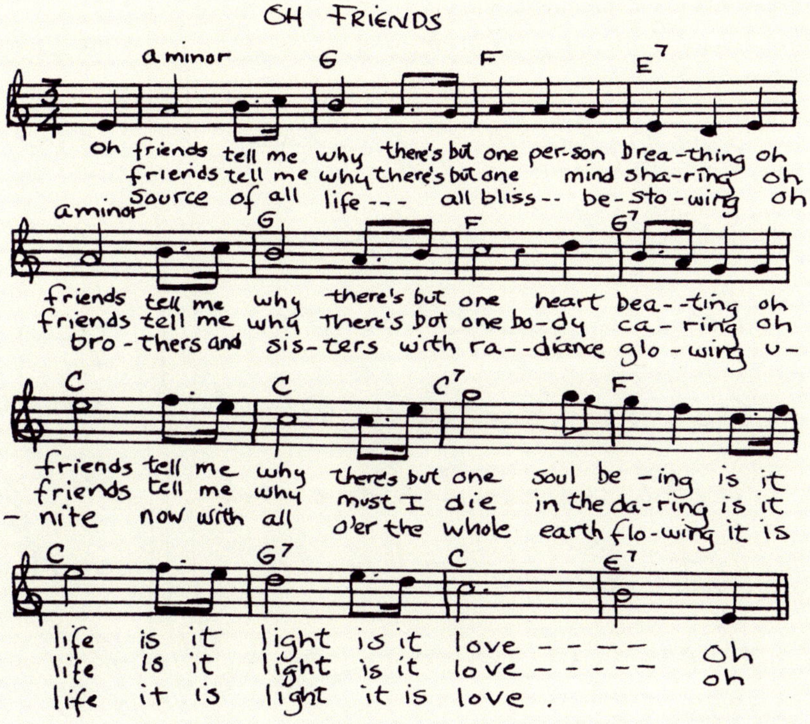

on open land, access is denied to no one, so that the land may select its own inhabitants. it is always a center of good vibrations. there are so few of these oases that they get overpopulated all too soon. we have great faith in the healing powers of total freedom. open sanctuaries could replace mental hospitals, jails, ghettoes, bars, boarding schools, ashrams, colleges & all haunts of lonely people. on open land you can camp out or build a simple shelter and do absolutely nothing until moved by god's will or hunger. but land should not have more people than it can support with ample privacy, water and public health.

Solitude is peace & quiet and freedom to do as you please. Alone in nature you find her rhythms as your living pattern, an enlightening balm for the nerves.

Solitude is only depressing if done indoors all the time. days spent silently observing & listening to nature are full of unexpected delights.

health

talk to your body

tell your body to be well.
appreciate its services.
stroke the tender parts.
encourage strength & grace
& relaxation.

talk to plants

tell them to grow green.
love them; touch them.
encourage beauty.

talk to nature

tell the universe
to be peaceful.
thank all things
for being.
serve anything
or anyone
in need.
encourage
all beings
to bliss.

everything has consciousness & can respond.

each person's needs are different. organically grown or wild foods which you gathered yourself from where you live are the most harmonious with your individual system, especially if you chew them well. not eating when not hungry is good for you. make all dietary changes gradually. food is made of sunlight & earth & should be gathered with love.

as the cells of your body absorb each substance you eat, they send messages of well-being or illness to your perceptive centers. to find your own perfect diet, eat very slowly

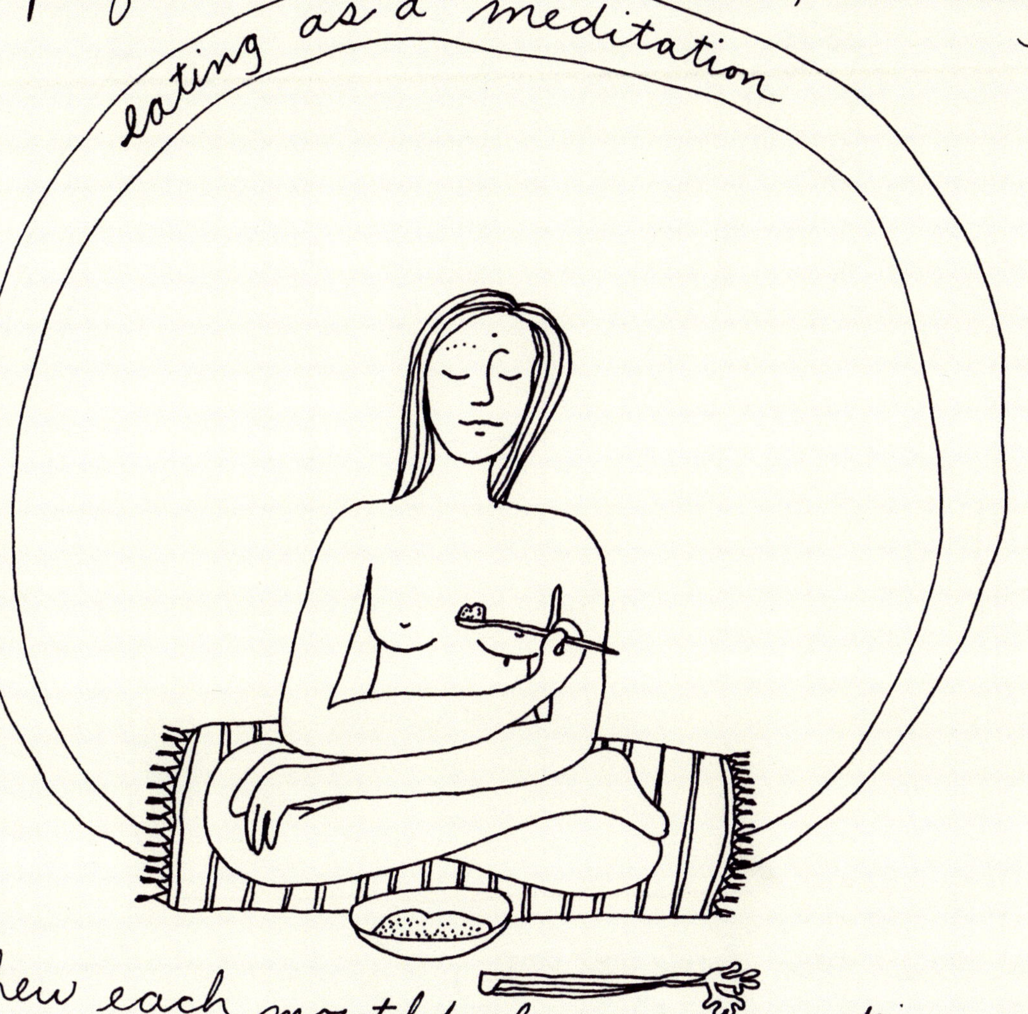

& chew each mouthful as many times as you can. feel the effects of each food on your stomach, intestines, and throughout your body. Our findings so far:

well-being foods	illness foods
kale, collards, chard, spinach, mustard greens, broccoli leaves (raw) raw garlic, wheat grass juice, carrot juice, papaya (with seeds), pineapple juice with brewer's yeast, raw goat milk yogurt, mangoes, oranges, passionfruit, pineapple, apples, berries, alfalfa sprouts, tofu (soy bean curd), avocado, kelp, herbal tea: alfalfa, mint, camomile, papaya leaf; rose hips, wheat germ oil, chia seeds, watercress, parsley	white sugar, white flour, anything containing insecticides, artificial sweeteners, hormone injections, preservatives; boiled vegetables, prepared supermarket foods, fermented tea, lard, margarine, (all solidified fats), instant foods, artificial or processed dairy foods, monosodium glutamate.

sickness is the body's way of eliminating toxic substances when it is overcome with them. vomiting, diarrhea, fever & sweat, pus-filled abcesses are all emergency cleansing measures. Industrially produced food, polluted air & water are the sources of the toxins. When the body is free of these toxins it never is sick. One can eliminate toxins through pure diet & fasting, but unless these changes are made gradually & with attention to the nutritional needs of the body, the elimination may be so rapid that it becomes sickness. rest up, keep warm, & drink lots of liquids.

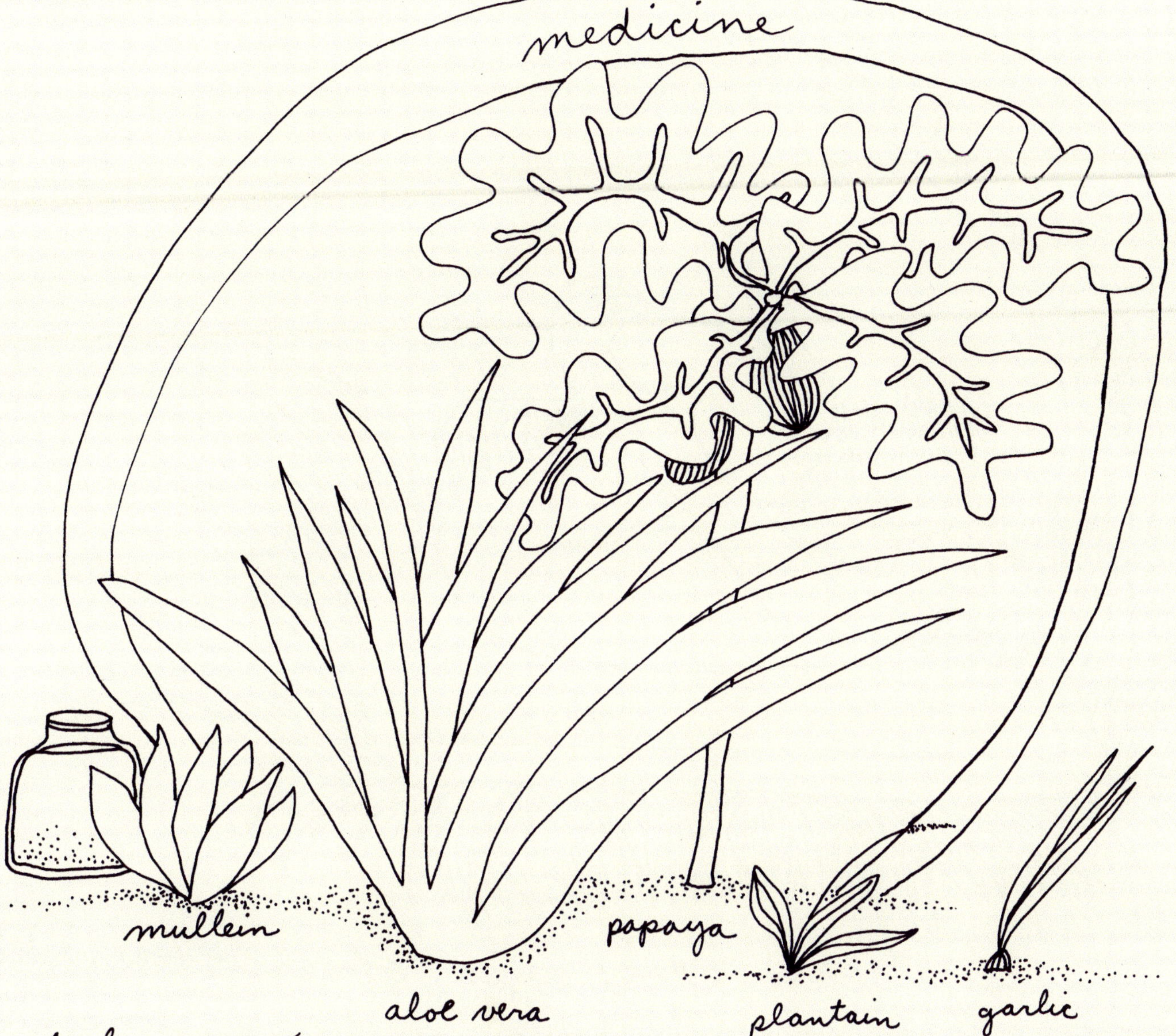

herbal cures (our favorites) raw garlic: suck clove for sore throat, oil clove & put in ear for earache, rub on insect stings, small infections & pimples; eat raw with meals to calm digestive tract, hold in mouth for tooth ache. rosemary tea: dandruff rinse. ginger tea: quiets farty intestines. peeled aloe vera leaves: rub on sunburn & other burns, mash & use as shampoo or eat to clean out colon. ginseng: stimulant. valerian root & passionflower herb: tranquilizer. powdered golden seal root or powdered myrrh: strong disinfectant for sores on skin or cuts. golden seal root powder (in capsules) for intestinal diseases.

preventative medicine learn which herbs cleanse different parts of the body. drink one each day. for example: plantain (for liver) one day, burdock (blood cleanser) the next day, cornsilk (for kidneys) next day, & so forth.

sun tea fill a clear glass jar with pure rain water add two teaspoons per quart of fresh herb or 1½ teaspoons per quart of dry herb. let stand in sunlight, shake well three times per day, for three days. strain & serve.

body meditations

sensing systems: close eyes. observe breathing. feel which muscles are tense. concentrate on your lungs, stomach, liver, all organs one at a time. you may even be able to feel your nervous system when you totally relax

fart cure: (above) clasp knees to chest. inhale. hold. slowly release. repeat until farts are gone.

hiccup cure: breathe deeply. hold your inhale. slowly exhale. hold empty. slowly inhale. hold again: repeat three times.

inside information: look at your shit. whatever was undigested should be eliminated from your diet in the future. a perfect diet means a perfect shit—it doesn't require toilet paper.

shitting into a hole in the earth is good for you & for the trees. it should be buried at least one foot deep & never uphill from or near a water supply. a post hole digger (above) makes a small round deep hole in minutes. Squatting tones the leg, back & abdominal muscles. if you must use a flush toilet, squatting instead of sitting protects you from hemorrhoids & constipation. a swedish anaerobic unit converts shit & organic garbage into sterile odorless compost; flush toilets pollute waterways. while meditating on sunlight, you can fulfill your deepest earthy role as builder of the soil and gardener of galaxies.

all places where water springs from the earth are holy. hot springs especially are medicine for people. Some sulphur springs contain traces of arsenic, which in such small doses is hallucinogenic. But sitting in water hotter than your body temperature is naturally tranquilizing anyway. (don't scald your skin.) steam clouds make good meditation points. If you can alternate the hot baths with dips in an icy cold stream your skin & circulation get to pulsating. the sulphur is good for curing infections in the skin. be sure to oil your skin well after rinsing off the sulphur water. please don't pollute the springs with soap...

keeping silence
is a good way to stop
thinking and automatically saying things
and initiating useless discussions.
Silence absorbs thoughts and cures
attention addictions. Instead, the sounds
of nature permeate you and soon
you hear the AUM of the universe
that was always droning
behind everything but was
drowned out by thoughts. listen
to your breathing, your
pulse, the wind & rivers.

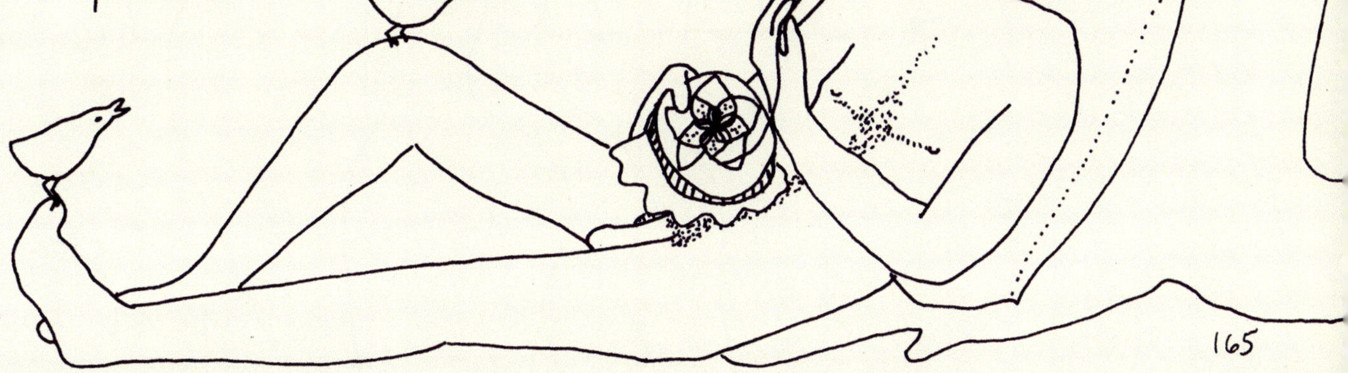

passion, joy, grief & anger really are emotional explosions. But there is a difference between self-expression and freaking out just to manipulate other people. consider the consequences for others, even for yourself before getting all wrought up about anything. usually it's not worth it. in solitude & silence & sunshine one may find one's cares begin to evaporate as trembling breaths slow down & tight assholes relax. try humming through your lips while sticking out your tongue.

worry

animals get their daily worrying done by chewing bones, grass, seaweed, hides, bark or leaves. people chew with their heads, gnawing old problems & imagining future ones. bring worry back to the body level by sucking your thumb or a pacifier or licorice root or some tough sundried apricots. or chomp on carrots, celery or cabbage. common worrybone substitutes: chewing gum, smoking, compulsive talking or eating.

often an unrelaxed state is due to nutritional deficiencies. Niacinamide (a B vitamin found in brewer's yeast), calcium (in milk) and vitamin A (carrot juice or a salad of raw spinach, collards, chard & mustard greens) are all essential to the nervous system. Also a noisy, electrical environment or lack of solitude may be the root of tension. In any case, the less you believe bad things will happen and the more you believe good things will happen, the more it works out that way. It's just a question of raising your bliss toleration level.

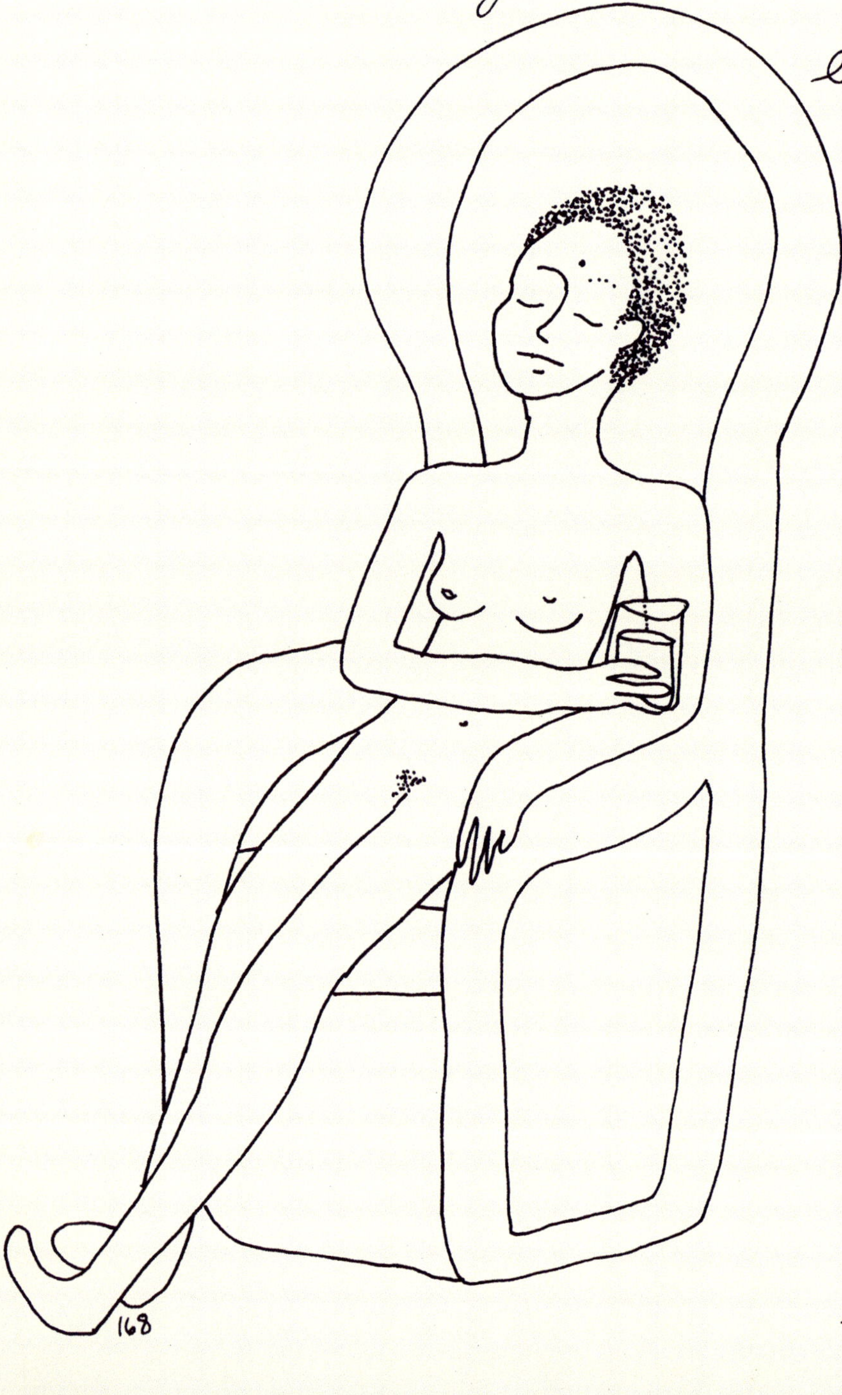

if you're in sore need of relaxation, take a long hot bath, drink some valerian root & passion flower herb tea, and lie on a bed or a hammock, expanding your asshole when you inhale and contracting it when you exhale, and if possible have someone you love massage you with oil.

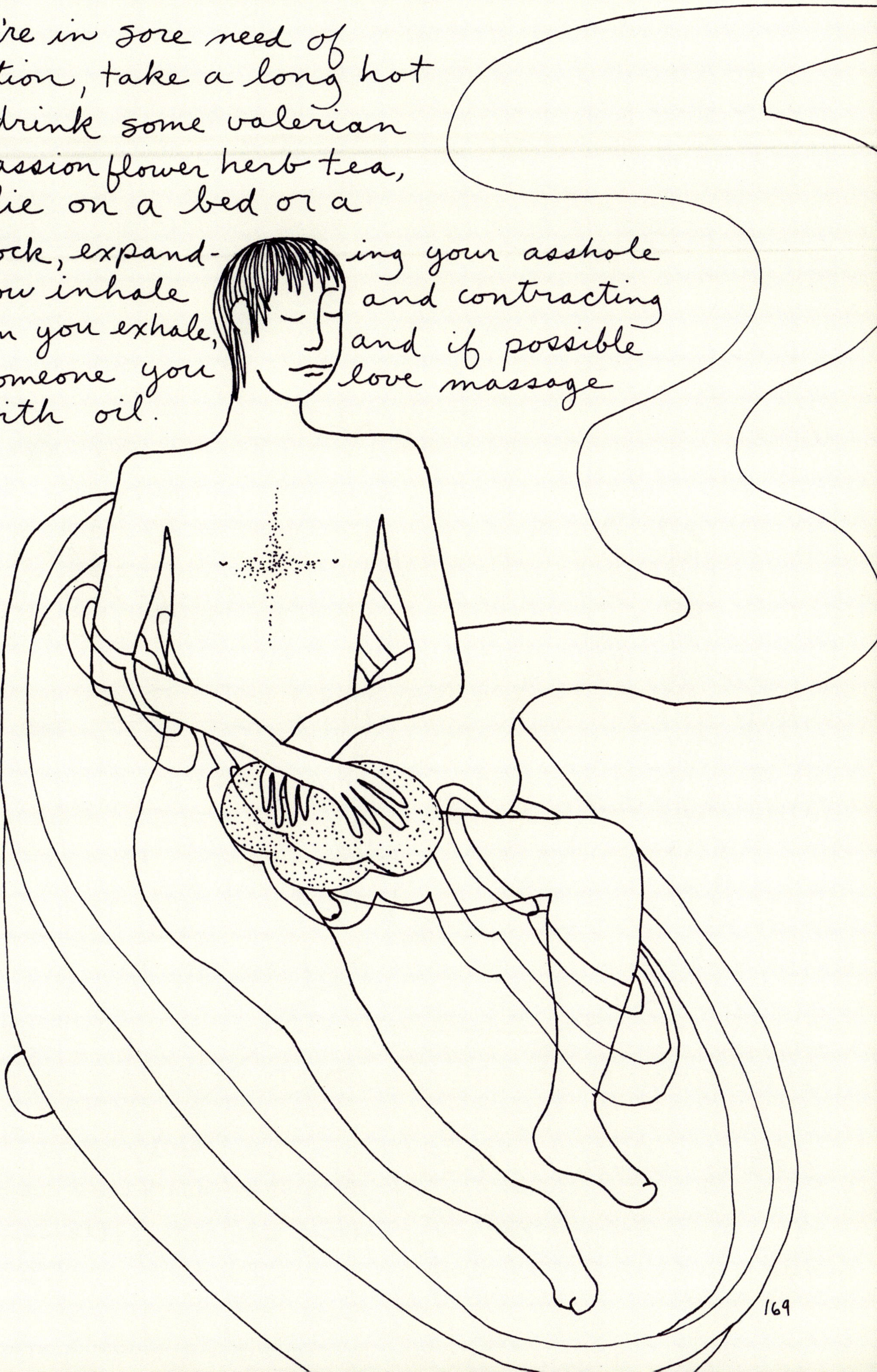

sitting around all day long brings great rewards in relaxation. No day spent sitting around is wasted unless it is done nervously.

if you are full of energy
and don't want to sit around
you can quiet your mind
by performing mindless
activity such as hiking
in wilderness, hatha yoga,
swimming, surfing,
dancing, gardening, cutting
firewood, gathering food,
singing, playing a
musical instrument,
potting, weaving,
sweeping, dusting,
washing things,
painting things
or taking a
bath.

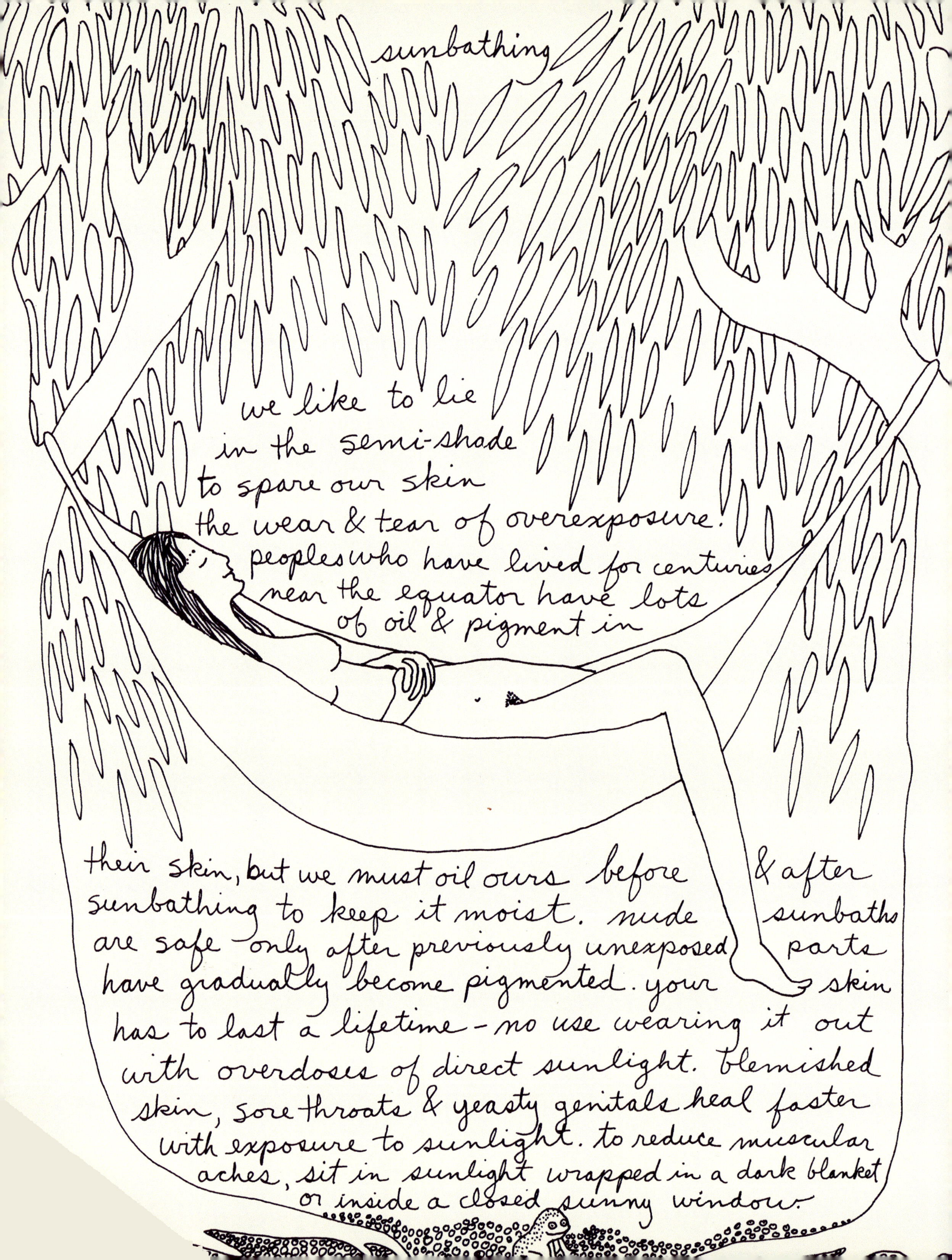

sunbathing

we like to lie in the semi-shade to spare our skin the wear & tear of overexposure! peoples who have lived for centuries near the equator have lots of oil & pigment in their skin, but we must oil ours before & after sunbathing to keep it moist. nude sunbaths are safe only after previously unexposed parts have gradually become pigmented. your skin has to last a lifetime - no use wearing it out with overdoses of direct sunlight. blemished skin, sore throats & yeasty genitals heal faster with exposure to sunlight. to reduce muscular aches, sit in sunlight wrapped in a dark blanket or inside a closed sunny window.

living naked

after one day of living outdoors naked your skin feels radiant and tingling. fresh air circulates on your skin instead of little pockets of stale air caught under layers of cloth. you become aware of your muscles, bones, skin, glands, organs as you never were when covered with clothes. after several days the temperature variation you are comfortable in will widen. the sunlight absorbed by your skin will give you both serenity and a decrease in appetite. but observe sunbathing precautions and avoid being in public view.

loose fitting clothes allow air to flow on the skin. Choose the most flowing & serene fabrics to surround your body. Use draw strings.

shoes the major nerves of the body terminate in the feet. Rigid bulky soles & elevated heels set the body off balance. Toes need air & room to spread out. on pavement wear rubber-soled shoes to absorb the shock. soft light-weight shoes allow the body to move with grace.

tightly bound hair creates tension in the body. jewelry puts extra weights & stress that inhibit free movement. Cosmetics pollute the skin & soaps dry it. Oils aid healing & elasticity. a massage of wheat germ oil after each bath makes the skin deliciously soft and smooth.

taking a sacrament

a sacrament is a soul medicine which opens your awareness of the here & now to the point where you may experience the void, the deity, or have visions. they have made available to anyone these experiences without years of discipline, and thus are a good cure for spiritual pride (holier-than-thou-ness). however most of them are presently illegal. if you take them, be far from civilization in a gentle natural environment and only with people you love who are serene. if someone is having a difficult experience with a sacrament, encourage the person to lie down on a comfortable bed and suck his thumb. give him some niacinamide tablets or brewer's yeast blended in pineapple juice.

being of the sunday.

one day a week (any day) put aside all survival activities and sit in nature observing the sunlight. prepare food the day before so nothing need be done by you or anyone else. by being of the sun one day a week you can live on the earth without getting caught up in the follies of civilization.

on being of the sunday
it's a good day to do some
of the things in this book.

when you have perfected your health, and your aura shows you to be without malice towards any being, and when you have totally relaxed & deconditioned to your god nature, and you know the healing powers of sunlight, herbs, good diet, silence, solitude, and all the preventative medicines, you are a sun healer. if you help others it will be without reward or payment and with the total consent of the recipient. although you may not practice medicine without a license, you may nurse the sick & unhappy after they have satisfied their need for professional help.

the more radiant sun healers & happy yogis of all kinds walk upon the earth, the closer we come to the golden age, when all beings exist in perfect harmony once again.

INDEX (also may be read as a poem)

hope you found
what you're
looking for.

APPENDIX A

Where to hear the music in this book

Song Title	composer/ lyricist	Page in this book	Album title	Track Number
Zither Meditation	Improvised by Alicia & Ramón	54-55	SFBOTS	4
Everything is Flowing	Improvised by Alicia & Ramón	64	SFBOTS	9
Gayatri	ancient sanskrit	70,134	SFBOTS	11
In the Morning	Alicia	72	MFLOTE	1, 17
Noon Chant	Ramón/ Alicia	76	SFBOTS	7
To The Sunlight	Ramón/ Alicia	78	N/A	
Solar Grace	Ramón/ Alicia	80	N/A	
Sunset Chant	Ramón	82	SFBOTS	2
Prayer for Night	Ramón/ Alicia	82	N/A	
Last Song of the Night	Alicia	84	MSFLOTE	16
4 Chants of the Moon's Phases	Ramón/ Alicia	89	N/A	
Spring Equinox Chant	Ramón/ Alicia	90	SFBOTS	3
Planting Day Ceremony	Ramón/ Alicia	93	SFBOTS	1
Sun is Forever a Lover	Alicia	97	SFBOTS	8
Summer Solstice Chant	Ramón/ Alicia	98	SFBOTS	5
Welcoming Chant	Ramón	103	SFBOTS	6
Fall Equinox Chant	Ramón/ Alicia	104	SFBOTS	10
Onward Ever Flow	Alicia	107	MSFLOTE	12
Thanksgiving Hymn	Alicia	109	MFLOTE	5
Winter Solstice Chant	Ramón/ Alicia	112	SFBOTS	15
Receiver~Believer	Alicia	113	N/A	
Green, Green Rain	Alicia	114	MSFLOTE	7
Eclipse Chant	Ramón/ Alicia	114	N/A	
Clouds Chant	Ramón/ Alicia	116	SFBOTS	12

Song Title	composer/ lyricist	Page in this book	Album Title	Track Number
Cloud Dispersing	Ramón	116	N/A	
Rain	Alicia	117	MFLOTE	3
Aditya Hridayam	Ancient Sanskrit	129	SFBOTS	14
Now	Alicia	129	N/A	
Oh Friends	Ramón	153	SFBOTS	13
Medley of solstice & equinox chants	Ramón/ Alicia	90, 98, 104 & 112	ABL-LIJ	3

Album Title Key

SFBOTS = Songs from Being of the Sun
MFLOTE = Music from Living on the Earth
MSFLOTE = More Songs from Living on the Earth
ABL-LIJ = Alicia Bay Laurel - Live in Japan

"Songs from Being of the Sun" began as a reel-to-reel tape recording Ramón made in 1973, of himself & Alicia performing songs from this book. 40 years later, Alicia found a cassette copy of the tape, and asked Ramón about the original reel-to-reel master. He located it in his archive at Mills College, where Maggi Payne, chair of the music department, made a digital copy, which Alicia had remastered in a studio & released as a CD.

alicia bay laurel books & music

living on the earth Bookworks (1970), Vintage Random House (1971-1980), Soshisha Ltd.-Japanese translation (1972-present), Villard Random House (2000-2003), Gibbs Smith, Publisher (2003-2010), Small Seed-Korean translation (2005-2008), Kachina Ediciones-Spanish translation (2017-present), Echo Point Books & Media (2020-present), Jiguhada-second Korean edition (2021 release).

being of the sun - with Ramón Sender Barayón Harper & Row (1973), Soshisha Ltd. - Japanese edition (1974-1980 & 2007-present), Echo Point Books & Media (2020-present)

earth time Random House (1972) art & poetry calendar

sylvie sunflower, family of families, happy day, cried the rainbow lady, full of light Harper & Row (1972), Soshisha Ltd. (1974) set of 3 children's books

earth mass Poetry by Joe Pintauro, illustration & design by Alicia Bay Laurel ☆ Harper & Row (1973)

Shakespeare's The Tempest Text by Michael Fleck Illustrations by Alicia Bay Laurel, New Age (1978)

MUSIC RECORDINGS by Alicia bay laurel

Produced by Alicia for Indigo With Stars, Inc.

Music from Living on the Earth (2000)

Spiritual & nature songs Alicia wrote while creating her first book

Living in Hawaii Style (2001)

8 original & 8 historic Hawaiian songs from Alicia's 30 years in the islands

What Living's All About (2006)

10 original & 2 historic jazz & blues tunes recorded with great LA session players

Beyond Living (2009)

Songs from a panoply of cultures that look at life, death & reincarnation, featuring many music legends

Living Through Young Eyes (2013)

8 solo fingerpicked guitar medleys that recall Alicia's childhood, teens & early 20s

Songs from Being of the Sun (2013)

Archival tape of Alicia & Ramón performing music from their book when it was first published, resurrected as a CD

More Songs from Living on the Earth (2015)

Songs about nature, spirit, love & revolution, written by Alicia when she made her first book, and recorded with many beloved guest musicians

Alicia Bay Laurel – Live in Japan (2018)

6 live recordings from Alicia's concert tours in Japan

Available at https://aliciabaylaurel.com (click on "Shop.") These lists were made in 2020. More recent releases will also be posted in the online store as they become available.

Ramón Sender Barayon: Books

Open Land Commune History

Home Free Home self-published (2017)
Morning Star & Wheeler's Communes Kindle (2016)
Home Free Home at: diggers.org/home_free.htm (1986)
Home Free: A History of Morning Star & Wheeler's self-published (1978)
The Morning Star Scrapbook Friends of Morning Star Press (1976)

Spanish Civil War & Ramón's Family

Death in Zamora with academic acclaim & commentary. Calm Unity Press, San Francisco (2019)
Muerte en Zamora second Spanish edition, with academic acclaim & commentary, Post Metropolis Editorial, Madrid (2017)
Death in Zamora (paperback), Calm Unity Press (2003)
Ein Tod in Zamora German edition, Kirchheim München (2000)
Muerte en Zamora first Spanish translation, with odious fascist footnotes by the publisher, added without notice to the author. Plaza & Janes, Barcelona (1990)
Death in Zamora first edition, University of New Mexico Press (1989)

Other

Being of the Sun with Alicia Bay Laurel ~ Harper & Row (1973), Soshisha Ltd. Japanese edition (1974-1980 & 2007-present), and Echo Point Books & Media (2020)
Zero Weather Family Publishing (1980), self-published (2020)
The Guide to Everwhere self-published on CD (1988)
Catalog of Prints, Drawings, Original Art (2009)

Ramón's books, continued

Planetary Sojourn self-published (2010)
Naked Close Up Intelligent Arts (2012)
Through self-published (2018)
Zero Summer Self-published (2020)

Recordings of Ramón

San Francisco Tape Music Center:

Performances of Ramón's compositions on a DVD included in David Bernstein's book *The San Francisco Tape Music Center*, University of California Press, 2008. Search YouTube by track number, plus the book title, for

video 4: "Tropical Fish Opera"
video 6: "Great Grandpa Lemuel's Death Rattle Reincarnation Blues"
video 8: "Koré"
video 11: "Desert Ambulance"

Also:

"Songs from Being of the Sun" with Alicia Bay Laurel (Indigo with Stars, 2013)
"Desert Ambulance" (Locust 70, 2005)
"Worldfood" (Locust 55, 2004)
"Gallivants & Garnishes" (Other Minds, 2016)
"Music from Mills"-"Audition" (Mills College, 1986)
"Desert Ambulance" 3-part podcast by Madrid's Museo Nacional Centro de Arte Reina Sofia (2020)
https://radio.museoreinasofia.es/en/desert-ambulance

appendix C: anatomy of the human eye

the human eye consists of a lens & a shutter that opens & shuts according to the amount of light shining on it—(the less the amount of light, the more it opens) and a screen of rods & cones on which the light is projected at the back of the eye. If these rods & cones are subjected to intense light, such as the direct rays of the sun, they can become permanently damaged; that is, you can be blinded. for this reason we urge our readers not to look directly at the sun or at brilliant reflections of the sun. If your eyes feel the slightest irritation when you are in the sunshine, close your eyes or go into the shade.

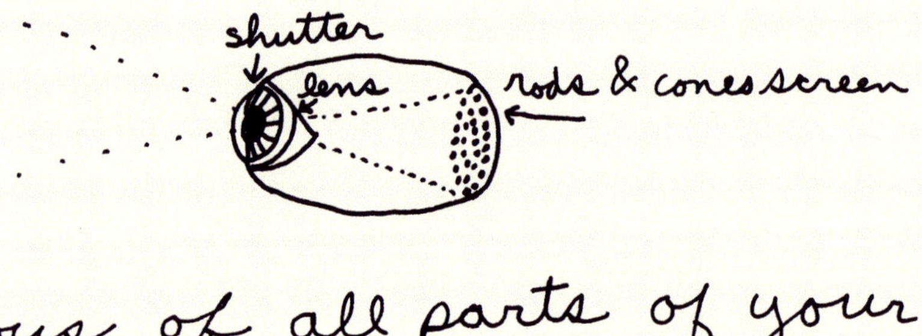

be conscious of all parts of your body and you'll enjoy them more.

this is not the end

because our learning of sun yoga will continue
as days of sunshine go by
and everyone & everything teaches us more.
thank you for visiting our book
hope you're having a good time
now and forever! ♡ Ramón & Alicia

about the authors

Ramón Sender Barayón, composer, musician, author, artist & philosopher, co-founded the San Francisco Tape Music Center in 1962, co-produced the San Francisco Trips Music Festival in 1966, and co-founded the Morningstar open land commune in 1967. He co-created *Being of the Sun*, published in 1973 and 2020. In 1982, Ramón wrote *A Death in Zamora* – the story of his family during the Spanish Civil War – published in 1989 and 2019. Ramón compiled *Home Free Home* (stories from the open land communes) published in 2017. In 2018, Spanish documentary filmmaker, Luis Olano, told Ramon's story in "Sender Barayon: Viaje Hacia la Luz."

alicia bay laurel, artist, author, singer/songwriter/guitarist/storyteller, wrote, illustrated & designed *living on the earth*, first published in 1970, followed by 8 other illustrated books, 6 of which she wrote or co-wrote. as of 2020, she has released 8 albums of original & historic songs, performed 12 solo concert tours in Japan, as well as concerts in Spain and in the USA. She also works as an illustrator & sells framed works. For more detail & updates, please see aliciabaylaurel.com/bio.